AF880668

I Can Do School Too!

NUMBERS

This Book Belongs to:

Copyright © 2022 by Sharise A. Caldwell

All rights reserved. No part of this book may be reproduced or transmitted in any form or by an electronic or mechanical means, including information storage and retrieval systems, without permission in writing from the publisher, excerpts by a reviewer who may quote brief passages in published reviews.

***Disclaimer** Most activities are purposed to use scissors in order to perform request. Parents are highly recommended to assist children as they cut these pages. We are not responsible for any accidents that may take place. It is under the parent own intellect to cut out the paper for the child or physically help them as they cut out the paper.*

Published by Caldwell Publishing Company
For more information, see www.Shariseantionette.com

Note You may share content in an appropriate setting exclusive to a learning environment, like study hall, classroom, library, small group discussion, bookclub, etc.

When shared on social media please tag us and feel free to use any combination of hashtags:

@ShariseAntionette
#MyBeautifulBookBoss
#MBBB
#ICanDoSchoolToo

ISBN: : 978-1-956318-12-8

Welcome to your brand new activity booklet.

Each activity will have clear instructions, will need adult supervision/help, and should only take 5-10 mins to complete.

We have included everything you will need to do the activities except for crayons, glue, scissors, and a cute toddler.

These activities are primarily focusing on numbers. The structure on each page is exactly the same with the same activities on repeat. This is so good for repetition and children who like doing the same things.

All instructions will need to be read and explained by adult with some assistance to complete each task.

Some activities will have more than one page involved but each page will be labeled to minimize confusion.

Feel free to do a new activity everyday or as much as the child would like!

Items you will need throughout this activity book:

*Scissors
*Glue stick
*Things to color with crayons, markers, colored pencils, paint, etc
*Sharpened Pencil
* A CUTE KID

NOTE to Parents

- Most activities in this book are for counting and coloring. Activity two is for cutting and pasting. Please assist child as they are developing their manipulative skills and strengthening their cutting abilities.

- If you decide to color the pages, please color **BEFORE** cutting and pasting to avoid trying to color over sticky glue, which at times will prevent the pages from being able to hold color.

- Enjoy your child as they have fun learning about their numbers. Help and encourage the child as much as they need when they are learning how to count from 1 to 25! They will soon master all of their numbers, coloring, and pasting skills!

Table of Contents

ONE
ONE Intro pg. 9
ONE Poster Board pg. 11
ONE Poster Objects pg. 13
ONE Example pg. 15

TWO
TWO Intro pg. 17
TWO Poster Board pg. 19
TWO Poster Objects pg. 21
TWO Example pg. 23

THREE
THREE Intro pg. 25
THREE Poster Board pg. 27
THREE Poster Objects pg. 29
THREE Example pg. 31

FOUR
FOUR Intro pg. 33
FOUR Poster Board pg. 35
FOUR Poster Objects pg. 37
FOUR Example pg. 39

FIVE
FIVE Intro pg. 41
FIVE Poster Board pg. 43
FIVE Poster Objects pg. 45
FIVE Example pg. 47

SIX
SIX Intro pg. 49
SIX Poster Board pg. 51
SIX Poster Objects pg. 53
SIX Example pg. 55

Table of Contents- Continued

SEVEN
- SEVEN Intro pg. 57
- SEVEN Poster Board pg. 59
- SEVEN Poster Objects pg. 61
- SEVEN Example pg. 63

EIGHT
- EIGHT Intro pg. 65
- EIGHT Poster Board pg. 67
- EIGHT Poster Objects pg. 69
- EIGHT Example pg. 71

NINE
- NINE Intro pg. 73
- NINE Poster Board pg. 75
- NINE Poster Objects pg. 77
- NINE Example pg. 79

TEN
- TEN Intro pg. 81
- TEN Poster Board pg. 83
- TEN Poster Objects pg. 85
- TEN Example pg. 87

ELEVEN
- ELEVEN Intro pg. 89
- ELEVEN Poster Board pg. 91
- ELEVEN Poster Objects pg. 93
- ELEVEN Example pg. 95

TWELVE
- TWELVE Intro pg. 97
- TWELVE Poster Board pg. 99
- TWELVE Poster Objects pg. 101
- TWELVE Example pg. 103

Table of Contents- Continued

THIRTEEN

THIRTEEN Intro pg. 105
THIRTEEN Poster Board pg. 107
THIRTEEN Poster Objects pg. 109
THIRTEEN Example pg. 111

FOURTEEN

FOURTEEN Intro pg. 113
FOURTEEN Poster Board pg. 115
FOURTEEN Poster Objects pg. 117
FOURTEEN Example pg. 119

FIFTEEN

FIFTEEN Intro pg. 121
FIFTEEN Poster Board pg. 123
FIFTEEN Poster Objects pg. 125
FIFTEEN Example pg. 127

SIXTEEN

SIXTEEN Intro pg. 129
SIXTEEN Poster Board pg. 131
SIXTEEN Poster Objects pg. 133
SIXTEEN Example pg. 135

SEVENTEEN

SEVENTEEN Intro pg. 137
SEVENTEEN Poster Board pg. 139
SEVENTEEN Poster Objects pg. 141
SEVENTEEN Example pg. 143

EIGHTEEN

EIGHTEEN Intro pg. 145
EIGHTEEN Poster Board pg. 147
EIGHTEEN Poster Objects pg. 149
EIGHTEEN Example pg. 151

NINETEEN

NINETEEN Intro pg. 153
NINETEEN Poster Board pg. 155

Table of Contents- Continued

NINETEEN Poster Objects pg. 157
NINETEEN Example pg. 159

TWENTY
TWENTY Intro pg. 161
TWENTY Poster Board pg. 163
TWENTY Poster Objects pg. 165
TWENTY Example pg. 167

TWENTY-ONE
TWENTY ONE Intro pg. 169
TWENTY ONE Poster Board pg. 171
TWENTY ONE Poster Objects pg. 173
TWENTY ONE Example pg. 175

TWENTY-TWO
TWENTY TWO Intro pg. 177
TWENTY TWO Poster Board pg. 179
TWENTY TWO Poster Objects pg. 181
TWENTY TWO Example pg. 183

TWENTY-THREE
TWENTY THREE Intro pg. 185
TWENTY THREE Poster Board pg. 187
TWENTY THREE Poster Objects pg. 189
TWENTY THREE Example pg. 191

TWENTY-FOUR
TWENTY FOUR Intro pg. 193
TWENTY FOUR Poster Board pg. 195
TWENTY FOUR Poster Objects pg. 197
TWENTY FOUR Example pg. 199

TWENTY-FIVE
TWENTY FIVE Intro pg. 201
TWENTY FIVE Poster Board pg. 203
TWENTY FIVE Poster Objects pg. 205
TWENTY FIVE Example pg. 207

"one"

1 one

one baby

One cute baby is ready to be colored. Can you count one with me? Say "ONE!" Good. Now let's say it again pointing at the number 1. Say " 1." One more time. Can you point to the baby or the word and say " One Baby." Awesome!

"One Poster"

On the next page you will find your objects to make this one's poster complete. Color your number 1, the word one, and the object representing one. Now cut them out and paste them right onto this page. Great Job!

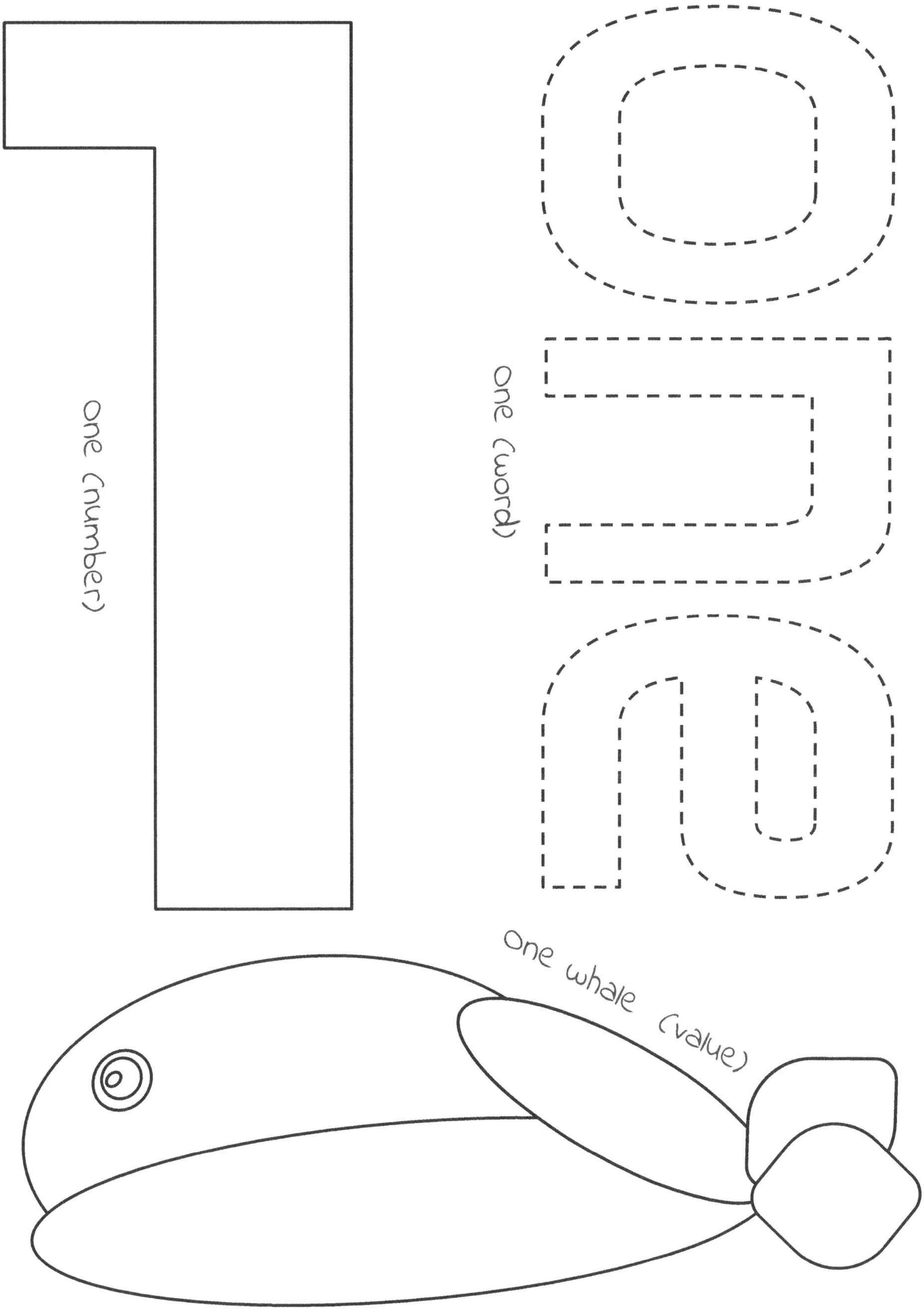
one (number)
one (word)
one whale (value)

"one"

one hairbow

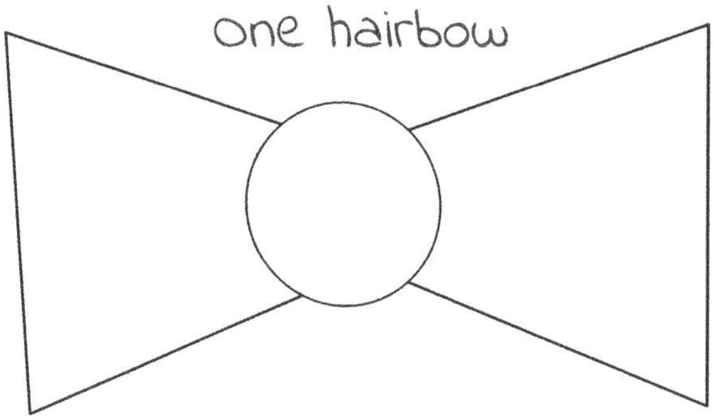

one bowtie

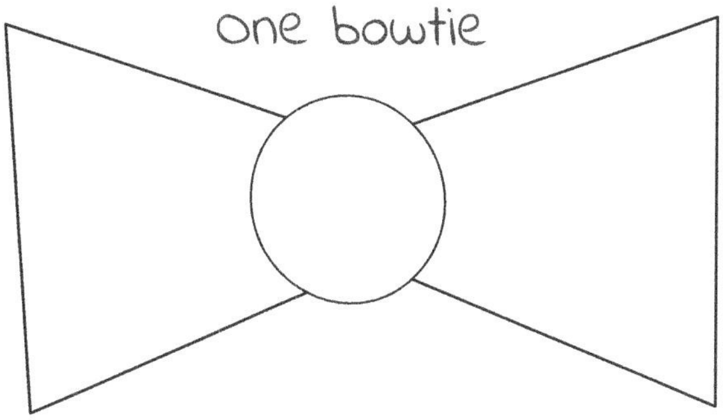

one

one cute girl

one handsome boy

This is your example sheet for the number one. You have one of each object. Let's point and say it together. "ONE HAIRBOW" "ONE BOWTIE." ONE CUTE GIRL." ONE HANDSOME BOY." Have fun coloring your pictures.

"two"

2 two

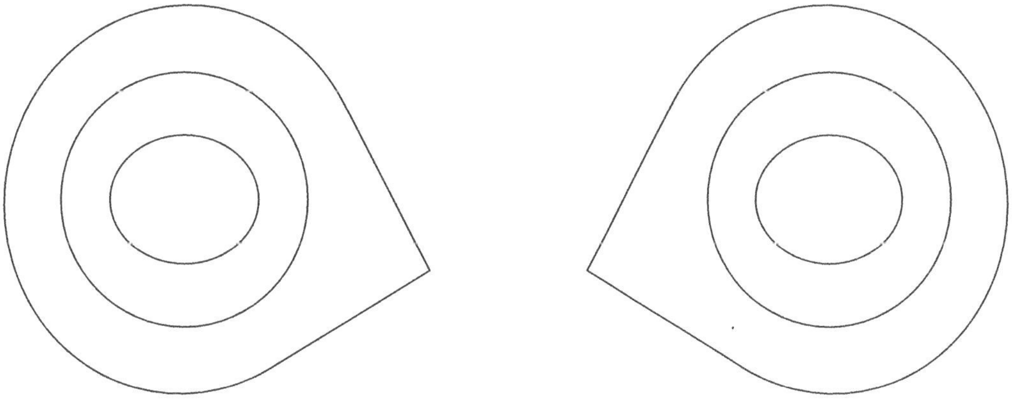

two eyes

What comes after the number ONE? That’s right, number 2. Let’s point to each object and say what it is. Point to the number 2 and say “TWO!” Point to the word two and say “TWO!” Point to the two eyes and count “ 1, 2!” Last, point to the word two eyes and say “TWO EYES!” Yay! You did it. Color away.

"Two Poster"

This is your number 2 poster sheet. On the next page you will have items all about number two. Color your objects and then paste them right on this page!

two (number)
two (word)
two hands (value)

"two"

On this page you have examples of the number 2. Let's point to each object and say what it is. Say, " TWO EYES!" Say, "TWO EARS!" Say, " TWO HANDS!" Now color your pictures.

"three"

3 three

three houses

What number comes after 2? That's right the number 3. Point to the number 3 and say "THREE!" Point to the word three and say "THREE!" Point to the three houses and count "1, 2, 3!" Last, point to the word three houses and say "THREE HOUSES!"

"Three Poster"

On the next page you will find your objects to make this three's poster complete. Color your number 3, the word three, and the object representing three. Now cut them out and paste them right onto this page. You're doing a wonderful job!

three (number)
three
three (word)
three trees
(value)

"three"

three staircases

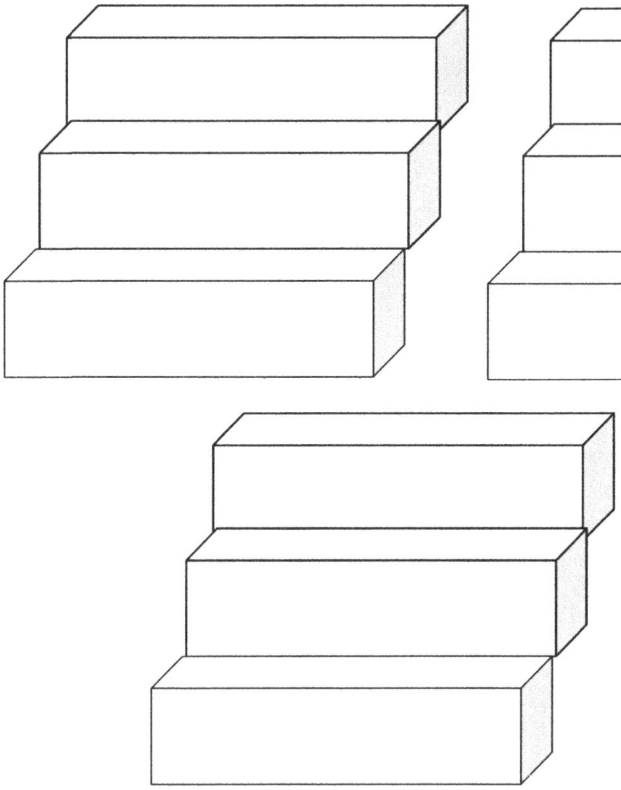

three doors

three windows

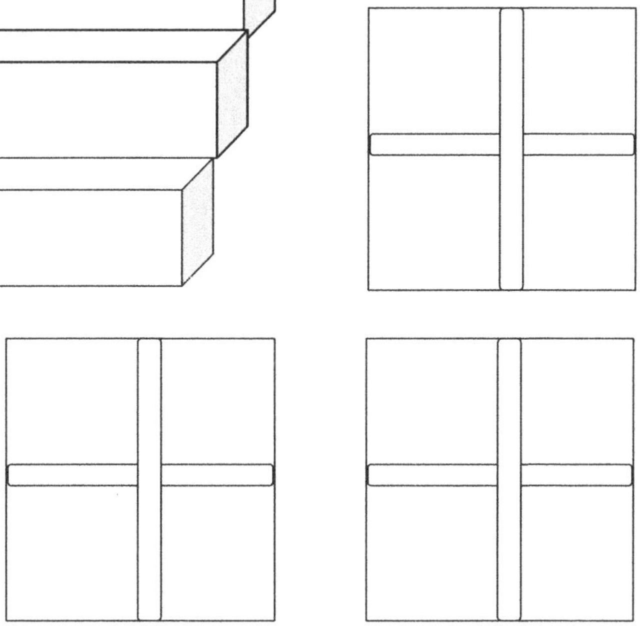

three

There are three examples of the number three. Let's point to each object and count them together. Point to the stairs and count, "1, 2, 3, stairs!" Point to the doors and count, " 1, 2, 3, doors!" Point to the 3 windows and count, " 1, 2, 3, windows!" Now color.

4 "four" four

four penguins

What comes after the number 3? You got it, 4! Time to point and count. Let's go. Point to the number 4 and say "FOUR!" Point to the word four and say "FOUR!" Point to the four penguins and count, " 1, 2, 3, 4, penguins!" Last, point to the word four penguins and say, " FOUR PENGUINS!"

"Four Poster"

This is your number four poster sheet. On the next page you have all about four.
Color each object, cut them out, and paste them on this page.

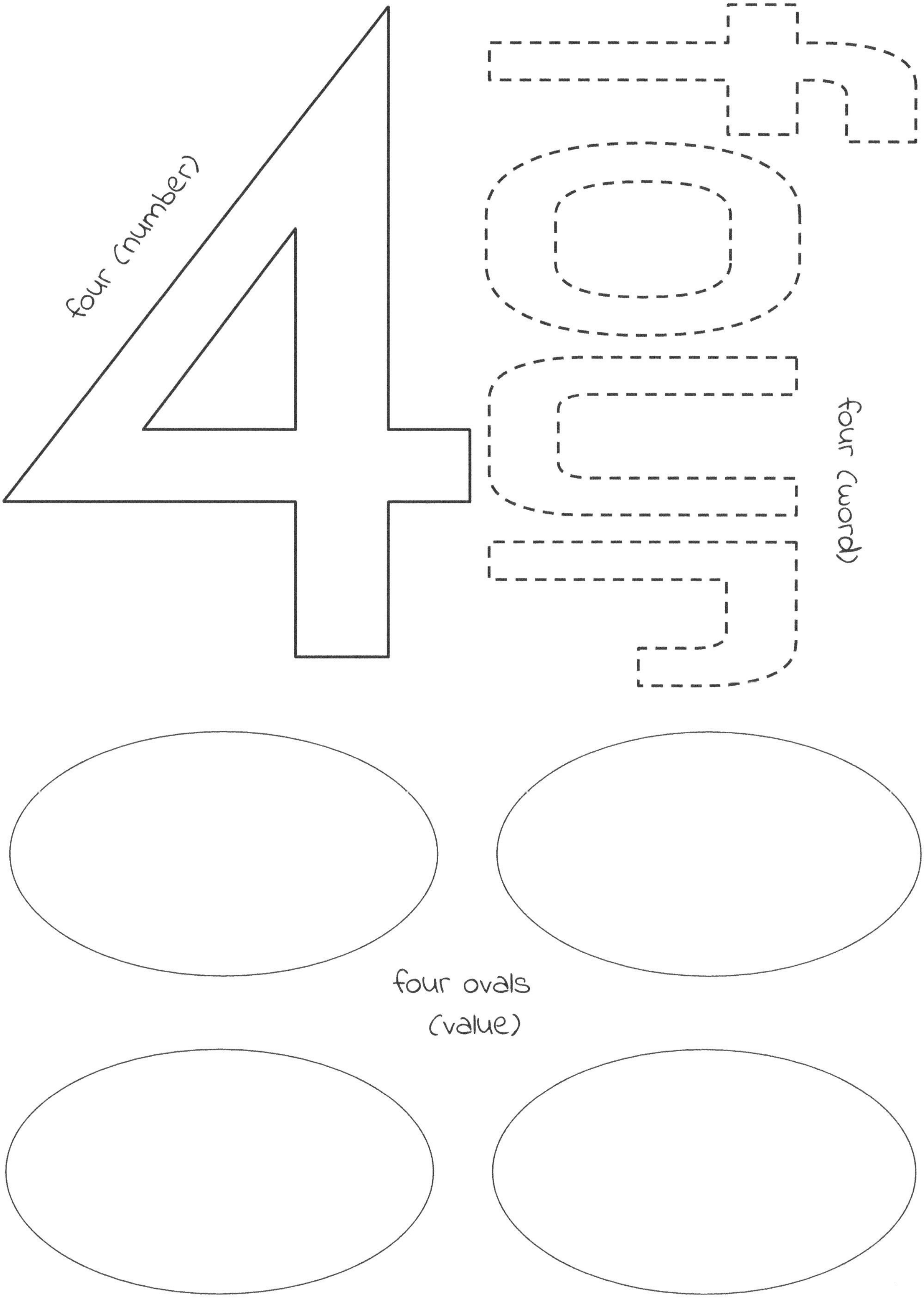
four (number)
four (word)
four ovals
(value)

"Four"

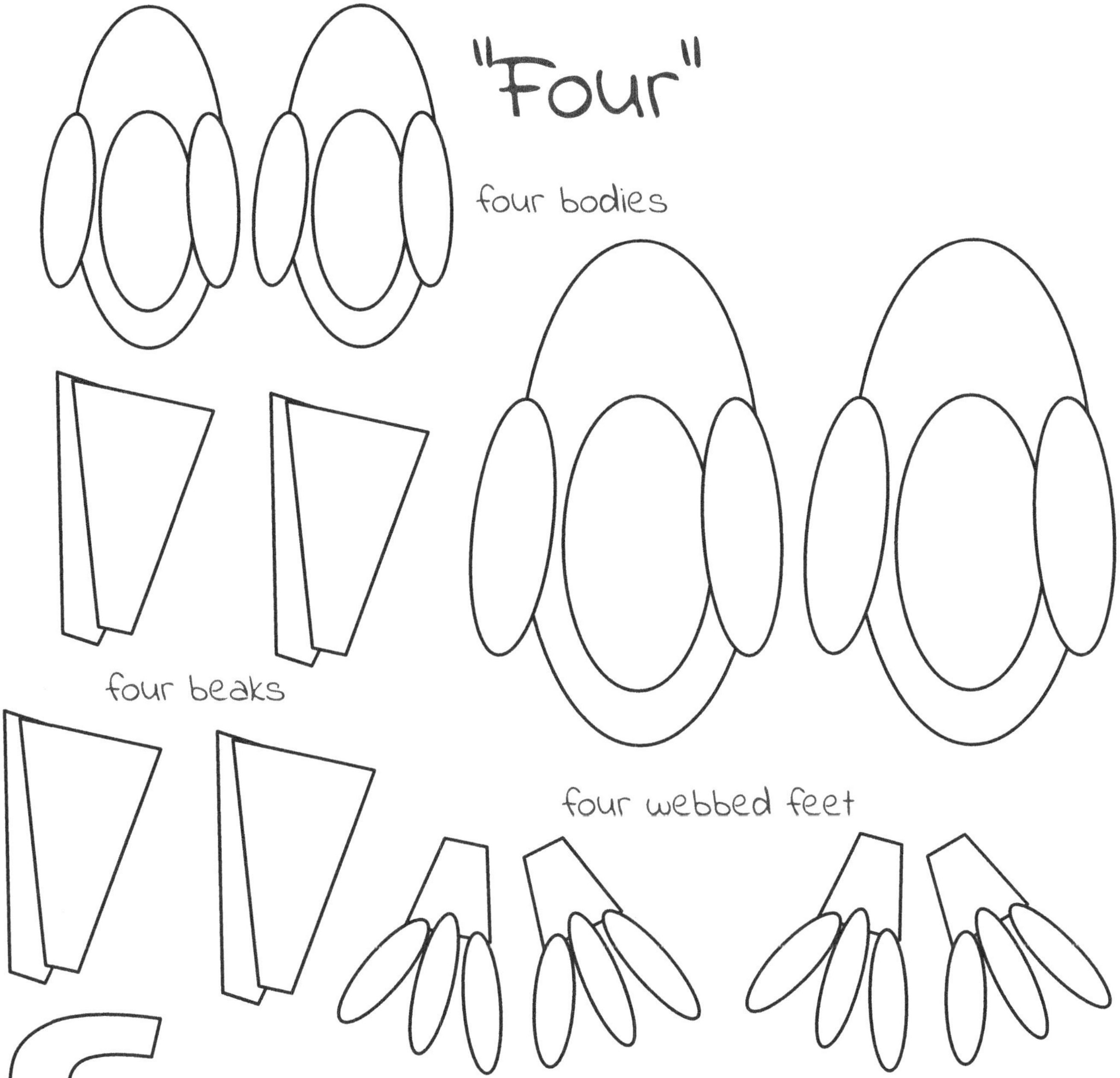

four

There are three examples of the number 4. Let's point to each group of objects and count them. Point to the four penguins and count, " 1, 2, 3, 4, penguins!" Point to the four beaks and count, " 1, 2, 3, 4, beaks!" Last, point to the four webbed feet and count, "1, 2, 3, 4, feet!"

"Five"

5 five

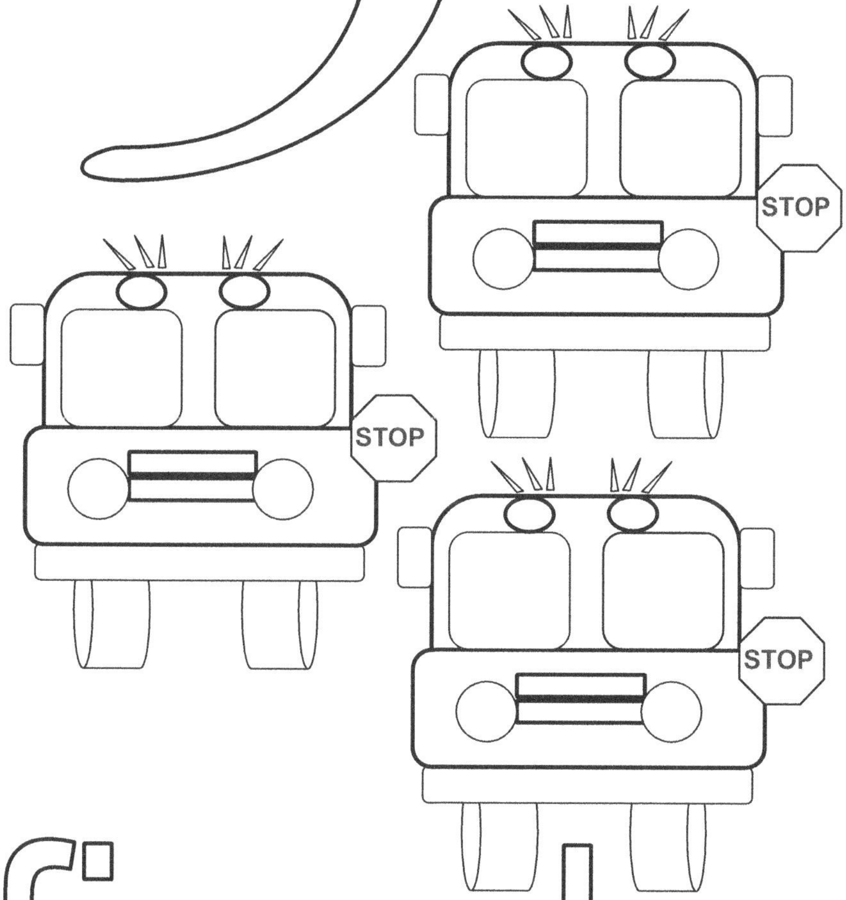

five buses

What comes after the number 4? 5! You're awesome! Point to the number 5 and say "FIVE!" Point to the word five and say "FIVE!" Point to each bus and count them with me, " 1, 2, 3, 4, 5, buses!" Last, point to the word five buses and say "FIVE BUSES!" Now color your picture nicely.

"Five Poster"

On the next page you have objects representing the number 5. Color all of your objects, cut them out, and paste them on this page. Now Go!

five (number)
five (word)
STOP
STOP
STOP
five stop signs (value)
STOP
STOP

"Five"

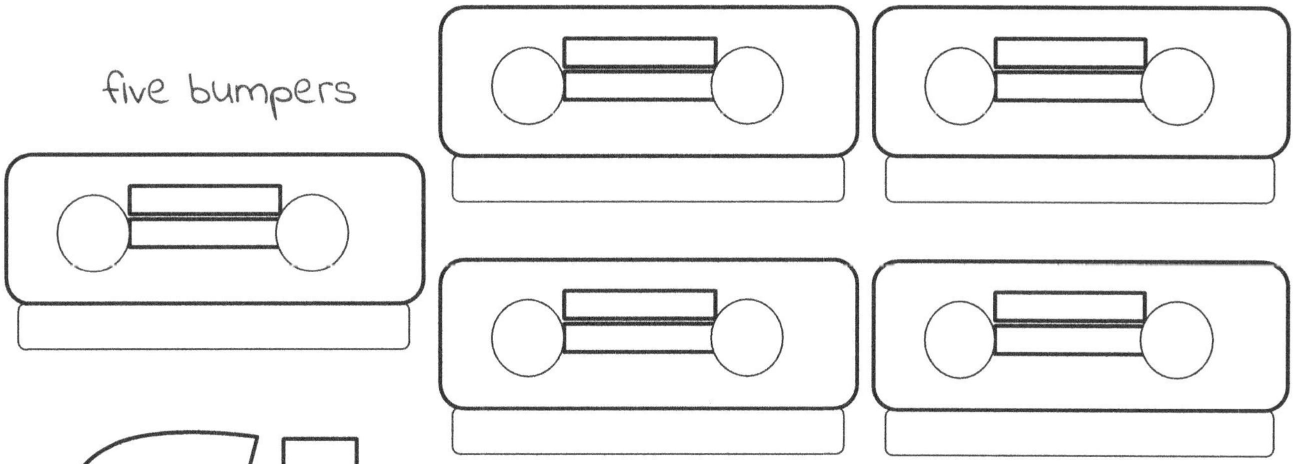

five

There are two examples of the number five. Let's point to each group of objects and count. Point to the tires and count, " 1, 2, 3, 4, 5, tires!" Point to the five bumpers and count, " 1, 2, 3, 4, 5, bumpers!" Color your awesome pictures.

6 "six" six

six lamps

What comes after the number 5? Did you get it right? 6! Point to the number 6 and say "SIX!" Point to the word six and say "SIX!" Let's count the lamps, say, " 1, 2, 3, 4, 5, 6, lamps!" Last, point to the word six lamps and say "SIX LAMPS!" Color your picture.

"Six Poster"

It's time to complete your number six poster sheet. On the next page, color all your objects, cut them out, and paste them on to this page. Keep up the good work!

Six round rugs (value)

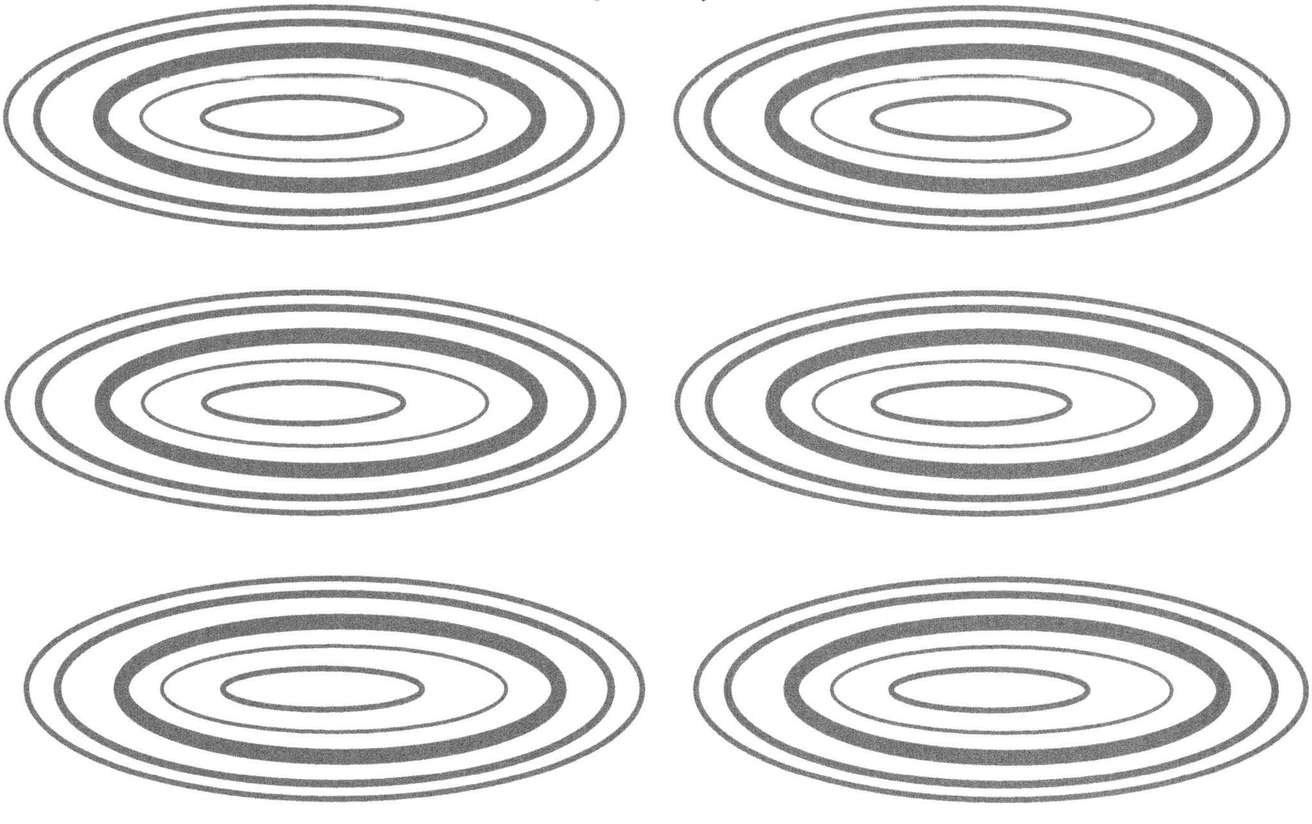

"Six"

There are two examples of the number six. We will count each group together. Point to the lightbulbs and count, " 1, 2, 3, 4, 5, 6, lights!" Now point to the lampshades and count, " 1, 2, 3, 4, 5, 6, shades!" Color your pictures to be amazing just like you!

7 "seven" seven

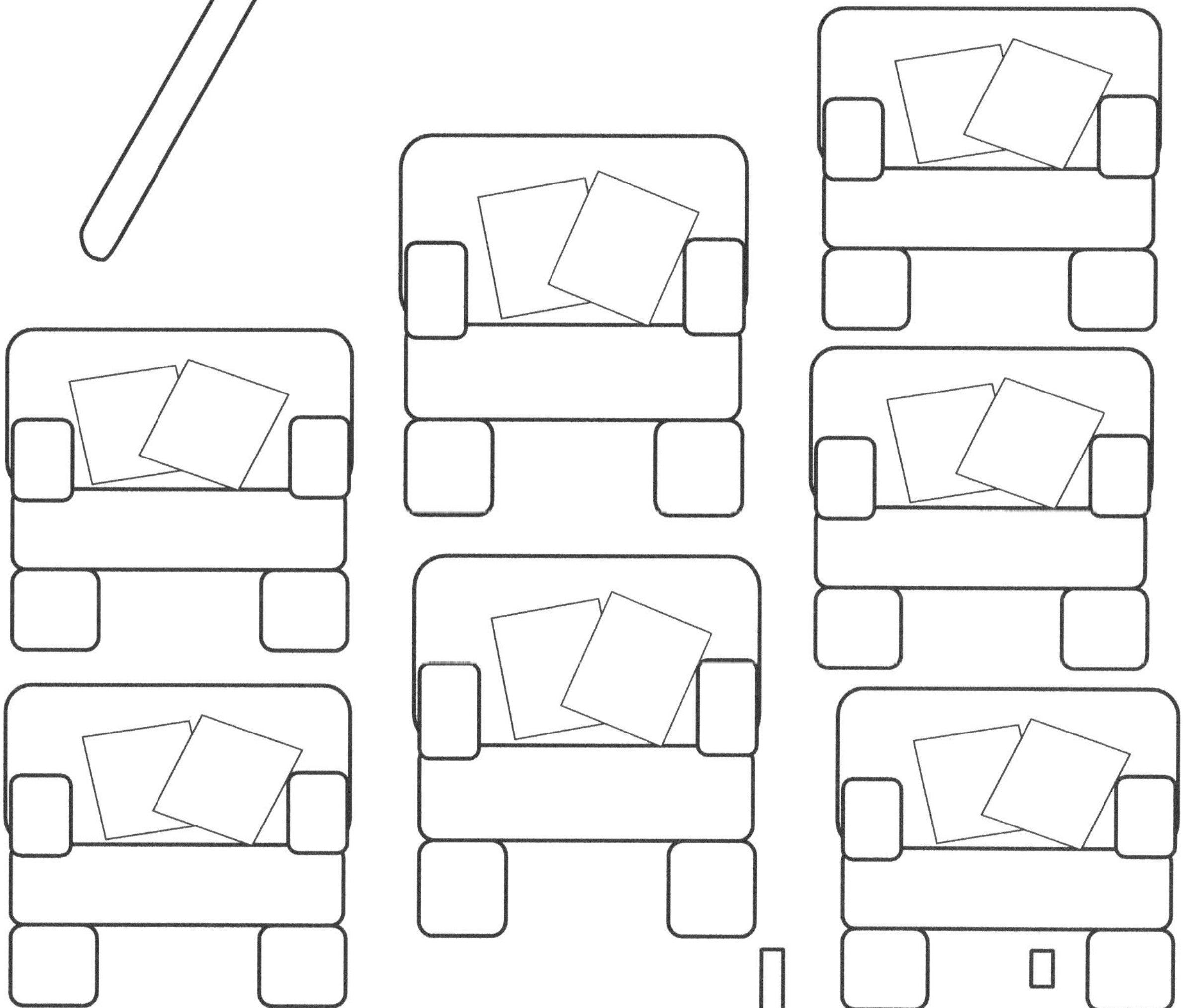

seven chairs

What comes after the number 6? Yes! 7. Point to the number 7 and say "SEVEN!" Point to to the word seven and say "SEVEN!" Point to the chairs and count them with me. " 1, 2, 3, 4, 5, 6, 7, chairs!" Point to the word seven chairs and say "SEVEN CHAIRS!" Coloring time!

"Seven Poster"

Fill out your seven's poster by coloring all the objects on the next page, cut them out, and paste them on to this page.

seven (number)
seven (word)
Seven wooden blocks (value)

There are seven cushions and pillows. Can you count them with me? Point to the cushions and count, " 1, 2, 3, 4, 5, 6, 7, cushions!" Point to the pillows and count, " 1, 2, 3, 4, 5, 6, 7, Pillows! Color your picture!

"eight"

8 eight

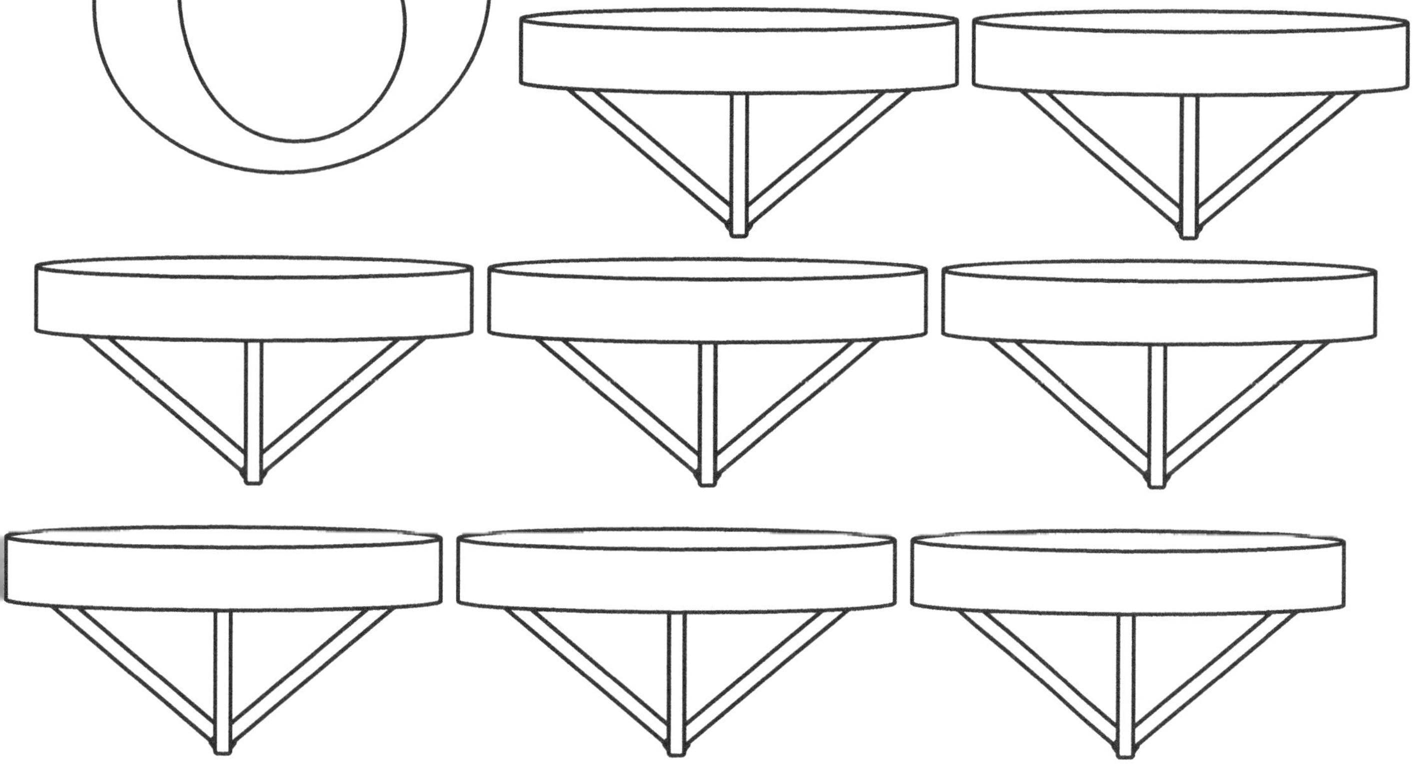

eight tables

What comes after the number 7? WOW! 8. Point to the number 8 and say "EIGHT!" Point to the word eight and say "EIGHT!" Let's count the tables, say, " 1, 2, 3, 4, 5, 6, 7, 8, tables!" Point to the word eight tables and say " EIGHT TABLES!" You did it! Yay! Color the picture!

"Eight Poster"

Color all of the pictures on the next page. Cut them out and paste them above. Your 8's poster is now complete!

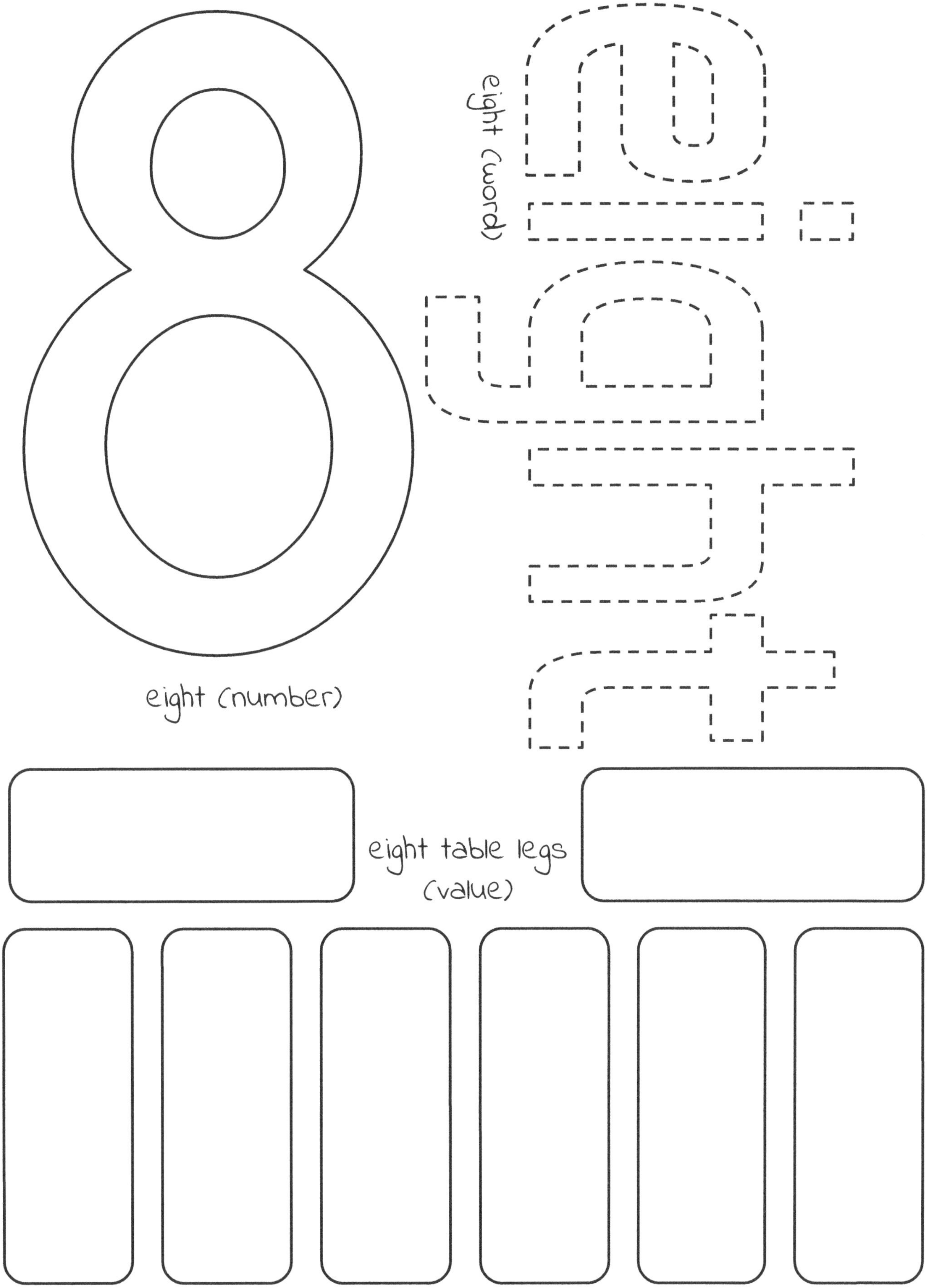
eight (word)
eight
eight (number)
eight table legs
(value)

"eight"

eight

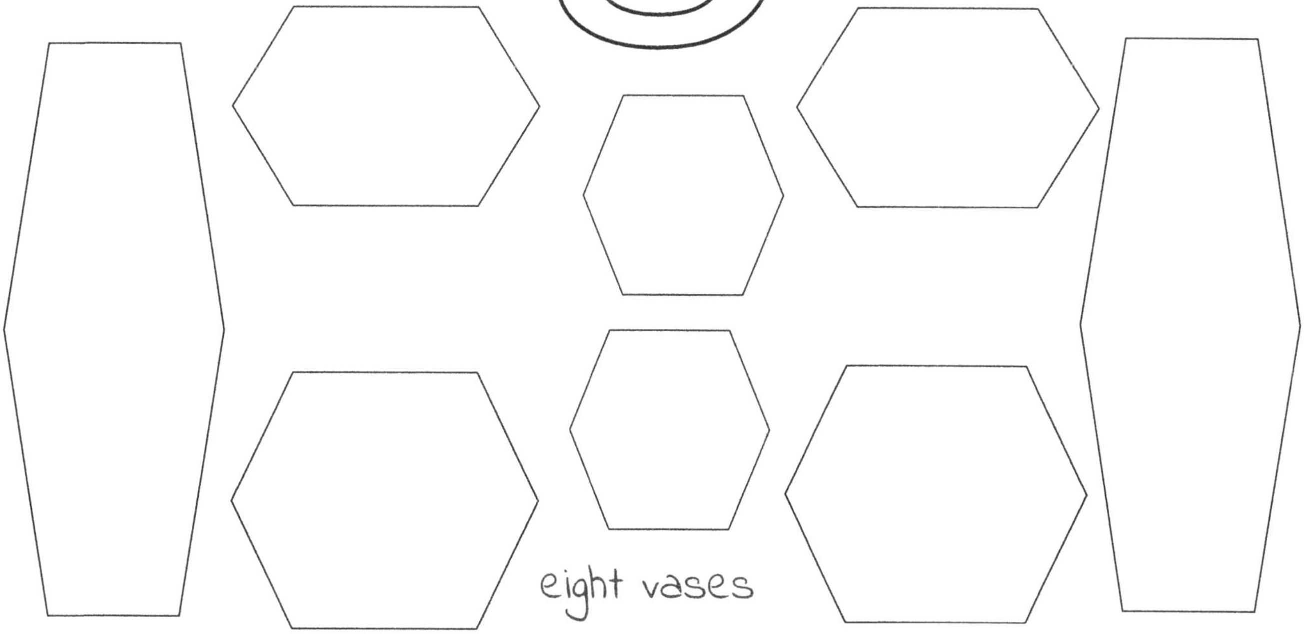

Point and count both groups of eight with me. Point to the table tops and count, " 1, 2, 3, 4, ,5 , 6, 7, 8 table tops! Point to the vases and count, " 1, 2, 3, 4, 5,6 ,7, 8, vases!" Color your beautiful picture!

What number comes after 8? You got it right again, 9! Point to the number 9 and say " NINE!" Point to the word nine and say, "NINE! Point to the books and count, " 1, 2, 3, 4, 5, 6, 7, 8, 9, books!" Point to the word nine books and say " NINE BOOKS!" Color everything!

"Nine Poster"

This worksheet is blank now but soon it will be covered. Go to the next page and you will find your objects. Color all the objects, cut them out, and paste them on this page.

nine (number)

nine (word)

nine labels
(value)

Have you ever had to help put your things away? Let's all of our book cases to make sure we have plenty of room for our toys. Point to the bookcases and count, " 1, 2, 3, 4, 5, 6, 7, 8, 9, bookcases!" Take your time coloring your shelves!

"ten"

10 ten

ten fire hydrants

Can you guess what comes after 9? 10! Correct! Fire hydrants are so important. They help us stay safe. Point to the number 10 and say "TEN!" Point to the word ten and say, "TEN!" Point and count the hydrants, "1, 2, 3, 4, 5, 6,7, 8, 9, 10, fire hydrants!" Point to the word ten fire hydrants and say " Ten Fire Hydrants!" Color your objects.

"Ten Poster"

What goes with a fire hydrant? Of course, a water hose. Color your pictures on the following page. Then, cute them all out and paste them above to finish the number 10's poster.

10

ten (number)

ten

ten (word)

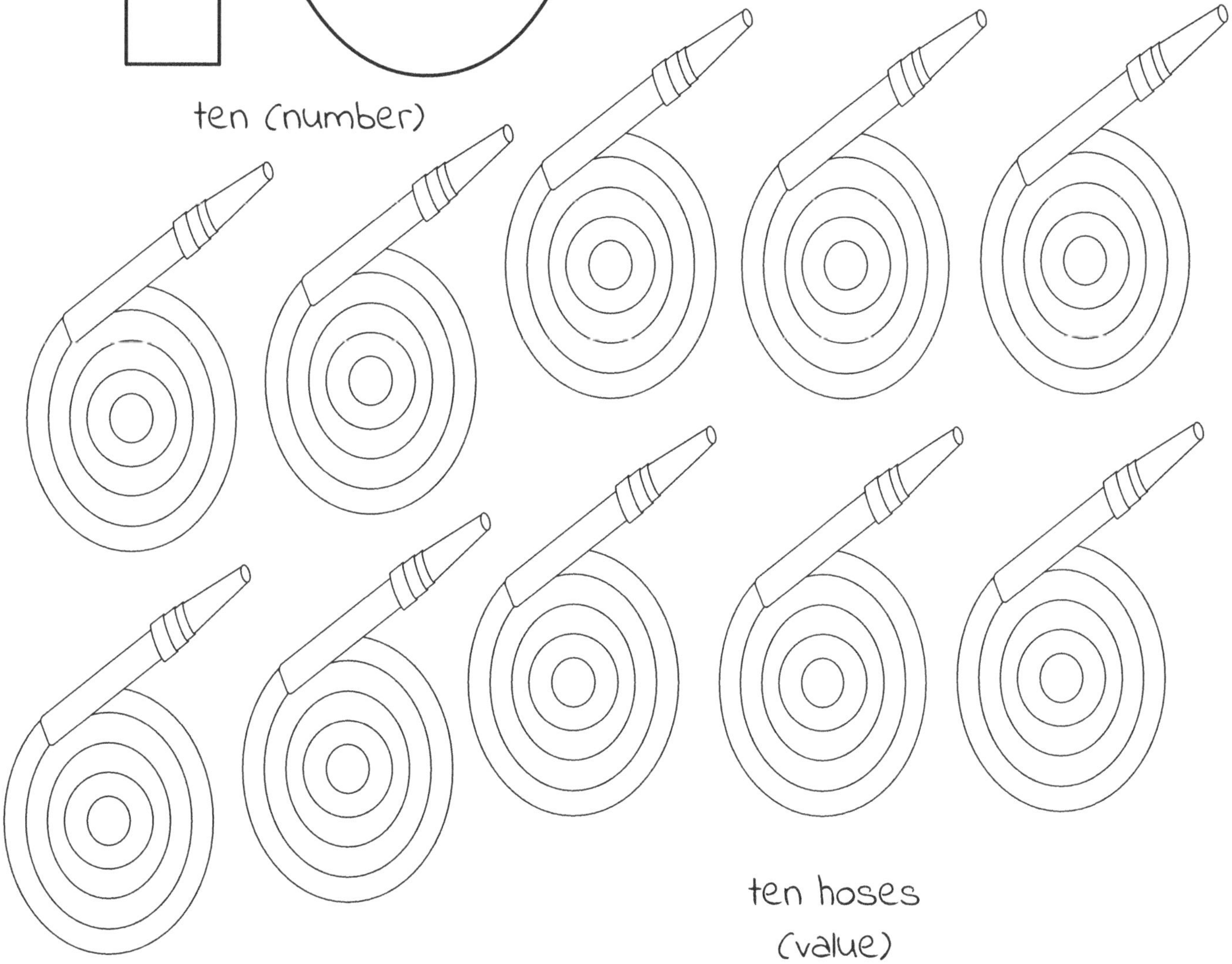

ten hoses
(value)

Being a fireman is so honorable. Let's count the objects. Point to the helmet and count " 1, 2, 3, 4, 5, 6, 7, 8, 9, 10, helmets!" Point to the badges and count, " 1, 2, 3, 4, 5, 6, 7, 8, 9, 10, badges!" Do a great job coloring this picture. Fantastic!

What comes after the number 10? It's 11! Awesome! Point to the number 11 and say, "ELEVEN!" Point to the word elven and say, "ELEVEN!" Point to the candles and count, " 1, 2, 3, 4, 5, 6, 7,8, 9, 10, 11, candles!" Point to the word eleven candles and say, "elevn candles. Have fun coloring!

"Eleven Poster"

Color your pictures about number 11 on the following page. Cut out all the objects and paste them to make a poster above.

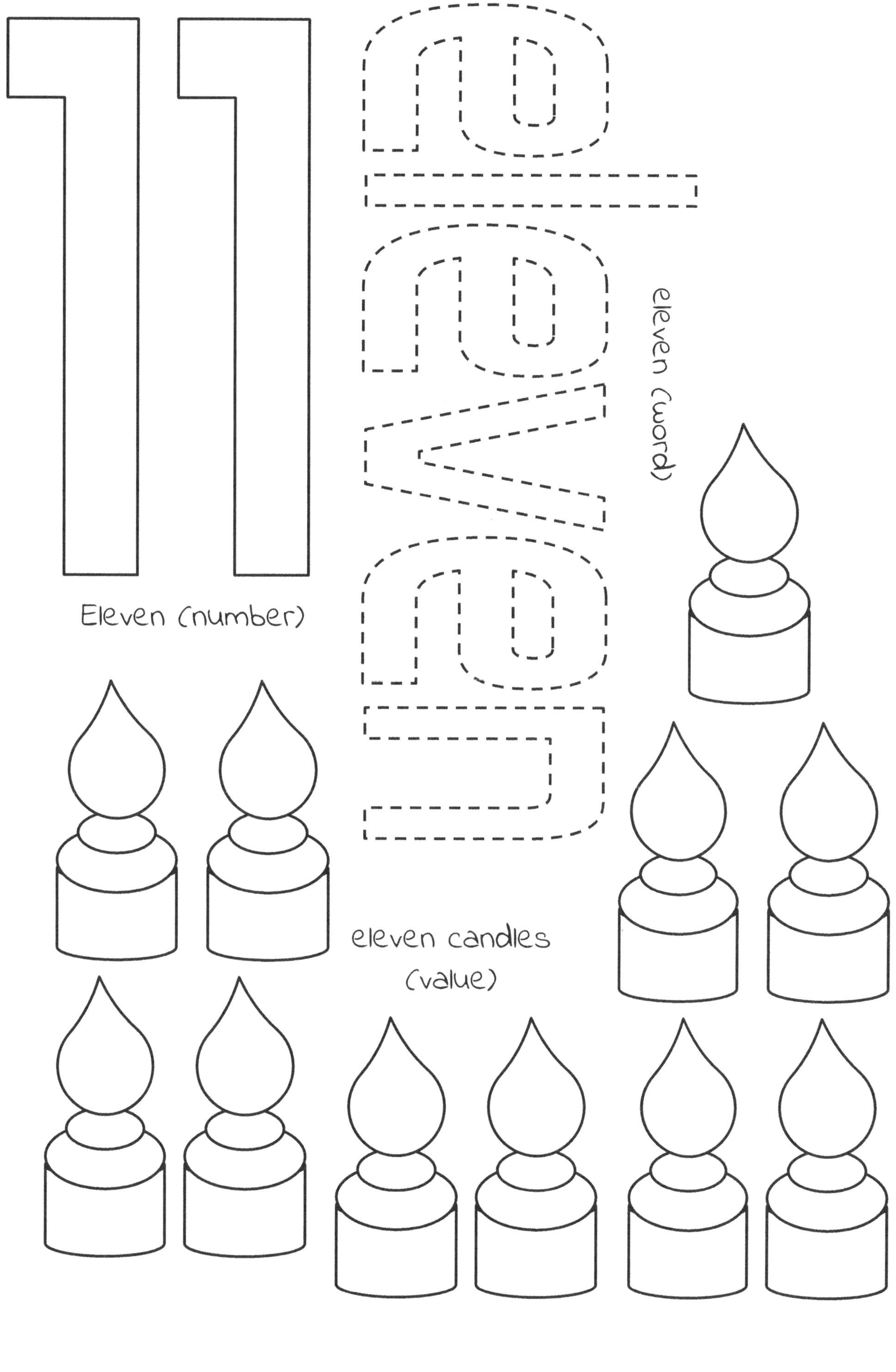
11
eleven
Eleven (number)
eleven (word)
eleven candles
(value)

What goes with candles? A match and flame! Point to the candles and count, " 1, 2, 3, 4, 5, 6, 7, 8, 9, 10, 11, matches!" Point to the flames and count, " 1, 2, 3, 4, 5, 6, 7, 8, 9, 10, 11, flames!" You are amazing! Color your picture!

"twelve"

12 twelve

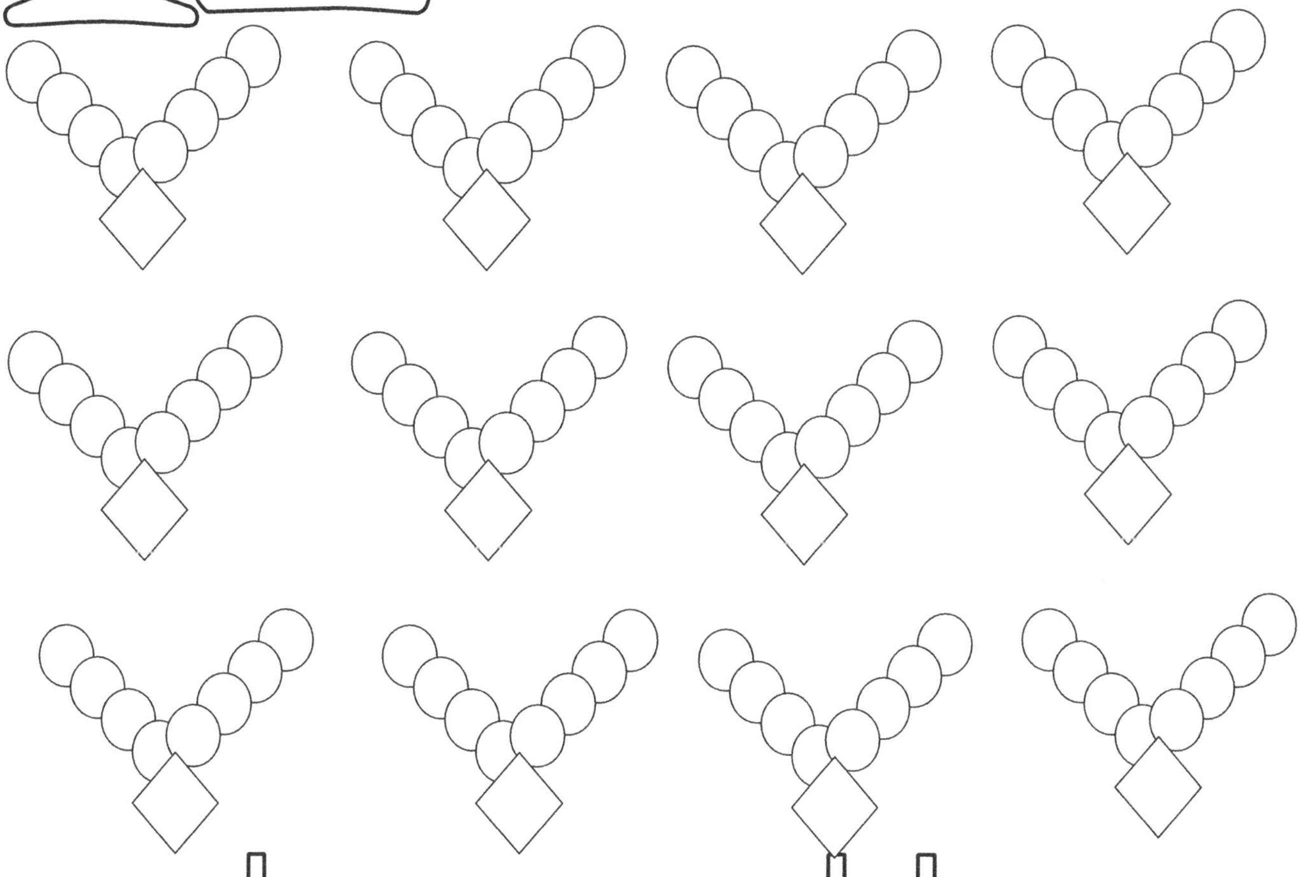

twelve necklaces

What number comes after 11? It's 12! Time to point and count. point to the number 12 and say, "TWELVE!" Point to the word twelve and say, "TWELVE" Point to the necklaces and count, " 1, 2, 3, 4, 5, 6, 7, 8, 9, 10, 11, 12, necklaces!" Point to the word twelve necklaces and say, " twelve necklaces!" Have fun coloring your picture!

"Twelve Poster"

Color your pictures all about number 12 on the next page. Cut out all the objects and paste them above to make a poster.

12

twelve (number)

twelve

twelve (word)

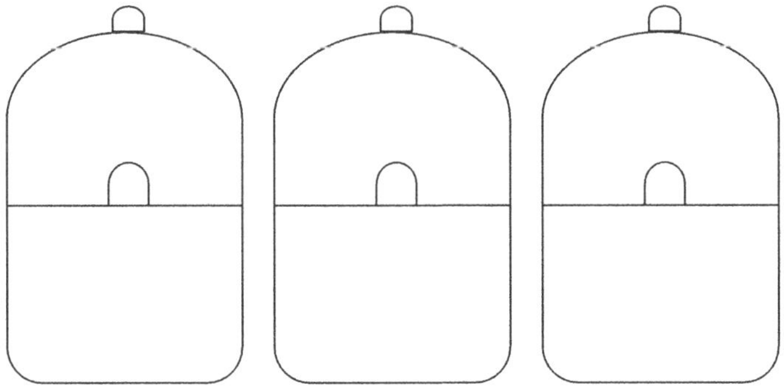

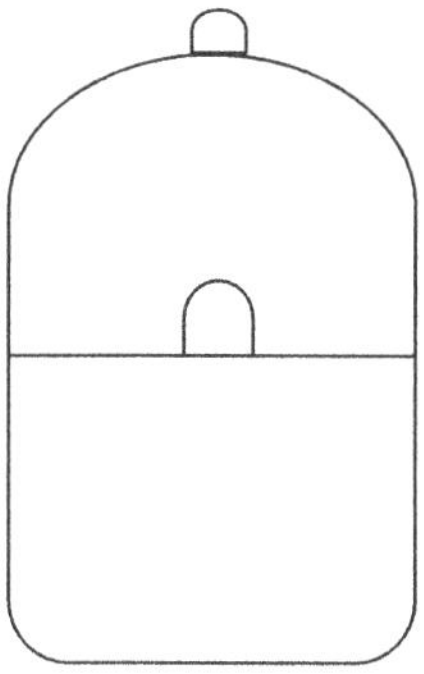

twelve treasure chest
(value)

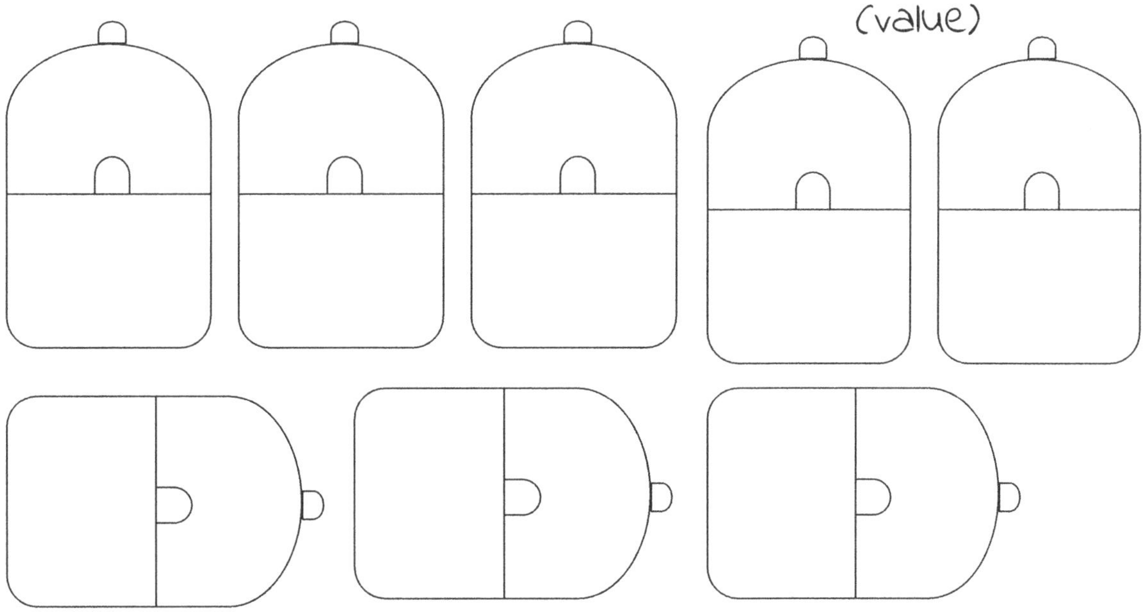

We can't have a treasure chest without the keys to get inside. Point to the keys and count, "1, 2, 3, 4, 5, 6, 7, 8, 9, 10, 11, 12, keys!" Point to the locks and count, "1, 2, 3, 4, 5, 6, 7, 8, 9, 10, 11, 12, locks!" Now color your picture!

"thirteen"

13 thirteen

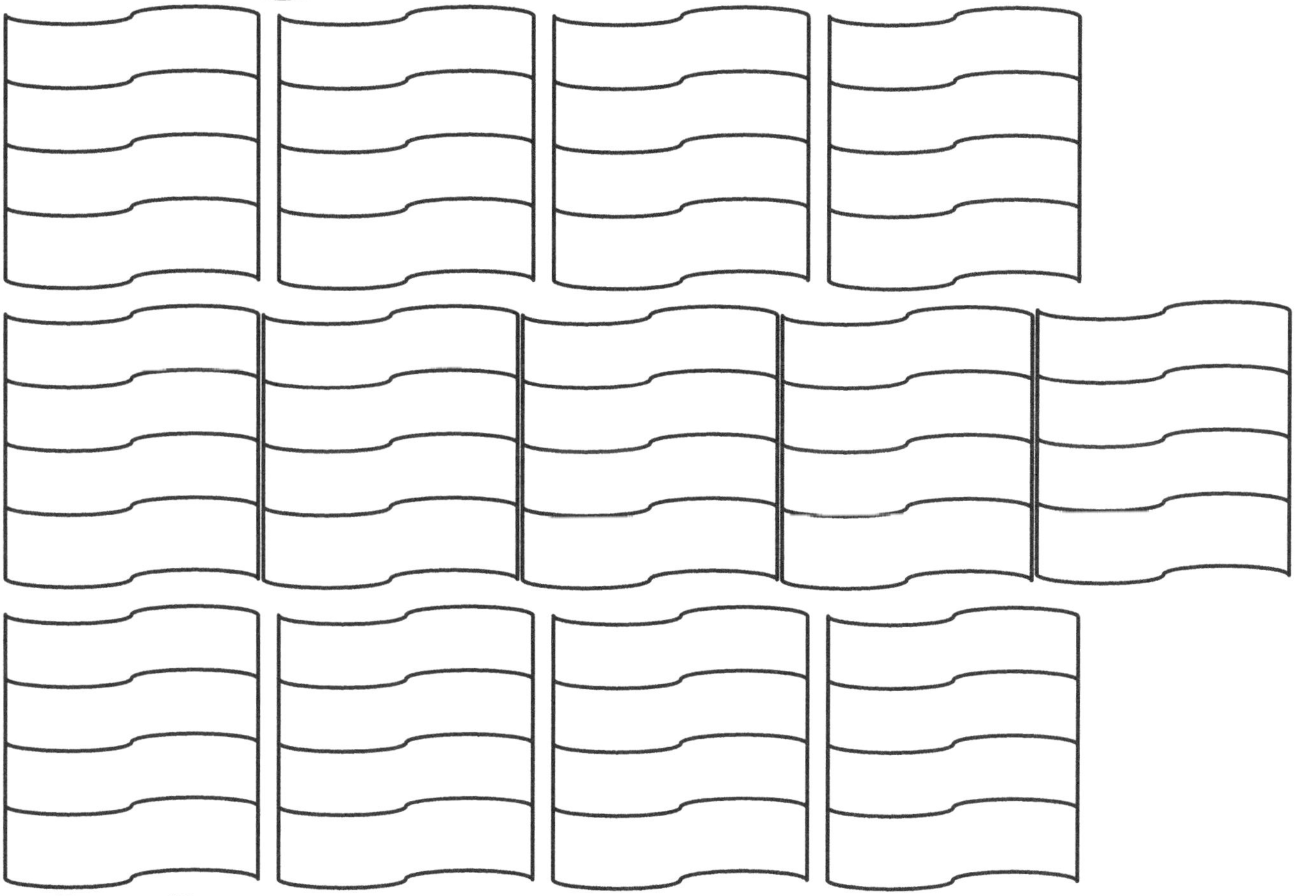

thirteen waves

Do you know what comes after 12? Yes, 13! Point to the number 13 and say, " THIRTEEN!" Point to the word thirteen and say, "THIRTEEN!" Point to the waves and count, " 1, 2, 3, 4,5 ,6 ,7 ,8 ,9 10, 11, 12, 13, waves! Point to the word thirteen waves and say "THIRTEEN WAVES!" Color your picture!

"Thirteen Poster"

Let's decorate this page by coloring the objects on the next page. After you color everything, cut them out, and paste them here! Now your poster sheet is complete!

thirteen (number)

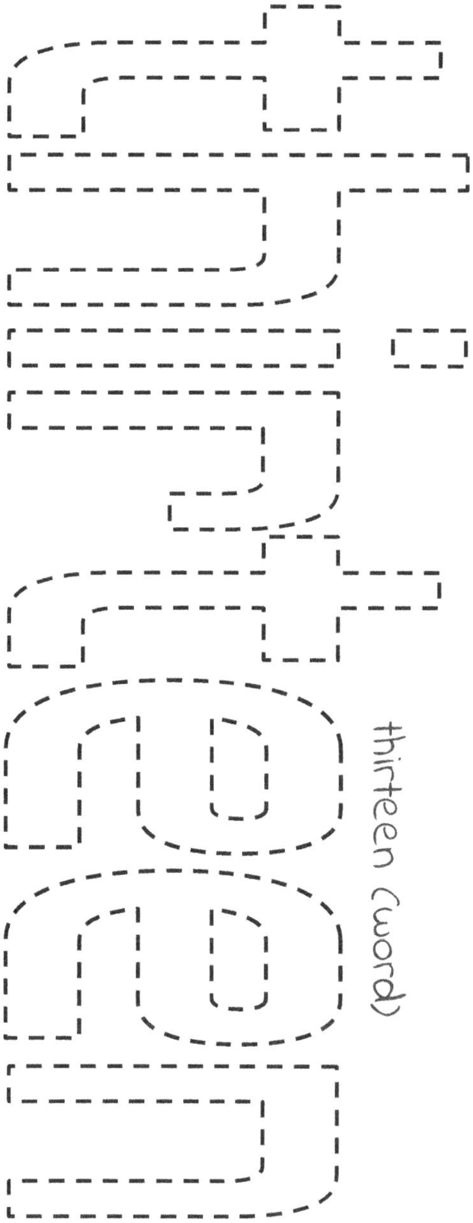

thirteen (word)

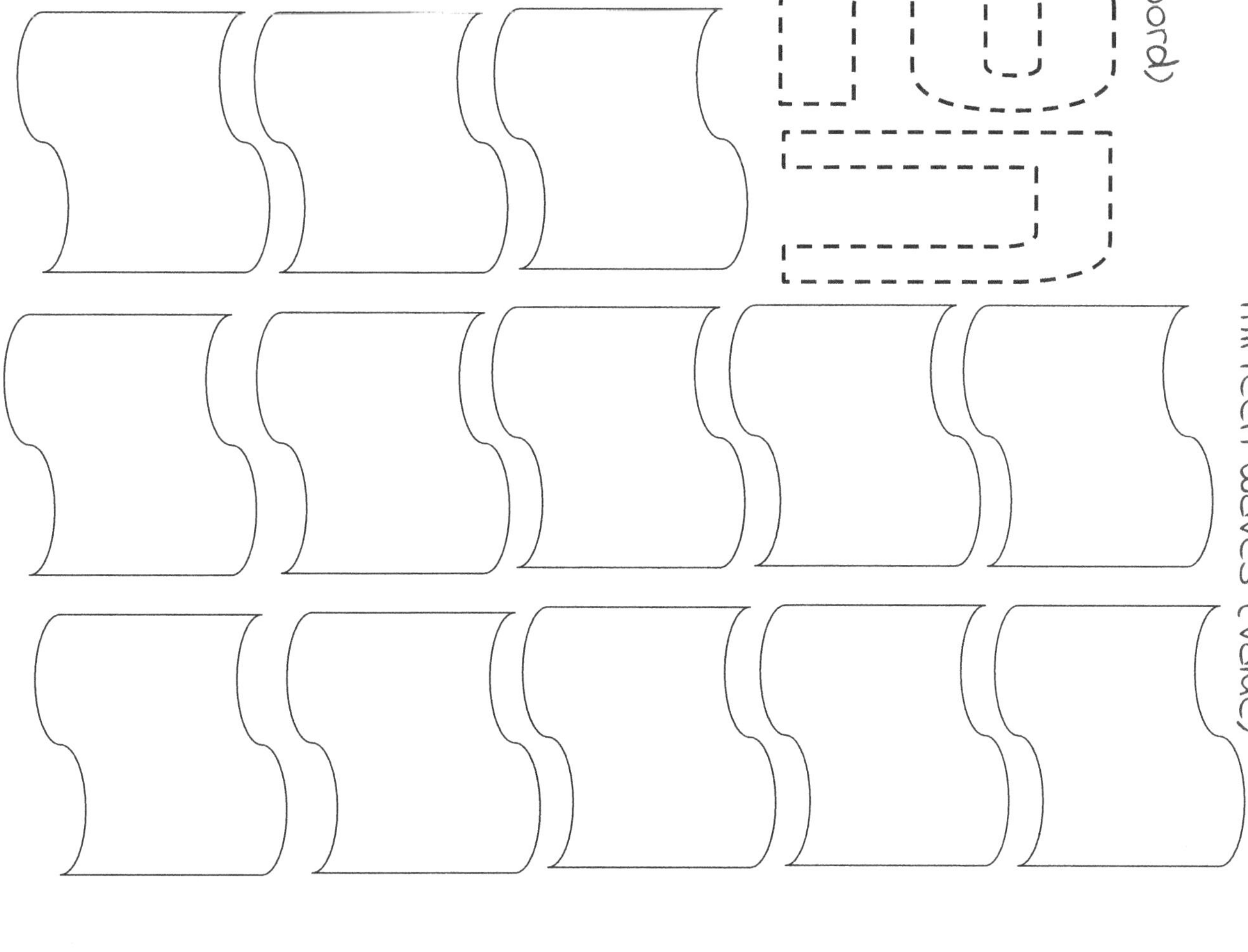

thirteen waves (value)

Warmer weather is so much fun especially when you are playing outside. Point to the bubbles and count with me, " 1, 2, 3, 4, 5, 6,7 ,8 , 9, 10, 11, 12, 13, bubbles!" Point to the wands and count, " 1, 2, 3, 4, 5, 6,7 ,8 , 9, 10, 11, 12, 13, wands!" Color your picture!

14 "fourteen" fourteen

fourteen sea shells

What comes after 13? 14! Let's point to the number 14 and say "FOURTEEN!" Point to the word fourteen and say, "FOURTEEN!" Point to the seashells and count, " 1, 2, 3, 4,5 , 6, 7,8, 9, 10, 11, 12, 13, 14, SEASHELLS!" Point to the word fourteen seashells and say, "FOURTEEN SEASHELLS!" Color the whole picture.

"Fourteen Poster"

Complete this poster worksheet by coloring all the objects on the next page. When you finish coloring, cut out the objects, and paste them on this page. What a good looking poster!

14

fourteen

fourteen (word)

fourteen (number)

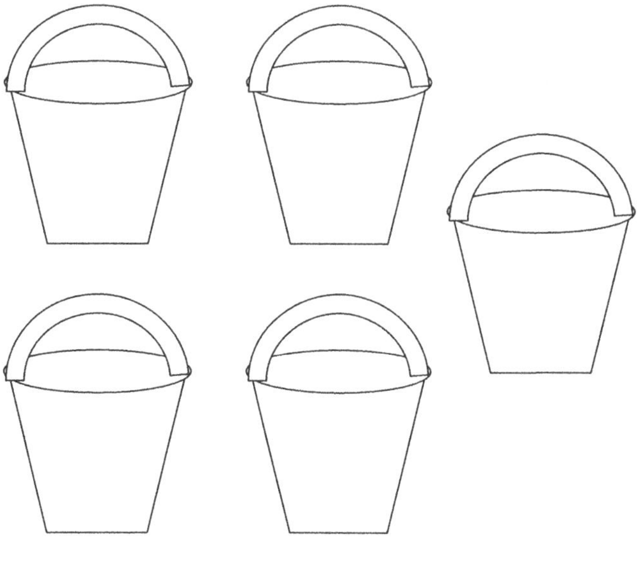

Fourteen pails (value)

Time to count up to 14. Let's go! Point to the plants and count, "1, 2, 3, 4, 5, 6, 7, 8, 9, 10, 11, 12, 13, 14, plants!" Point to the fish and count, "1, 2, 3, 4, 5, 6, 7, 8, 9, 10, 11, 12, 13, 14, fish!" Color your amazing picture!

15

"Fifteen"

fifteen

fifteen trees

What comes after the number 14? 15! You're so smart! Point to the number 15 and say, "FIFTEEN!" Point to the word fifteen and say, "FIFTEEN!" Point to the trees and count with me, " 1, 2, 3, 4, 5, 6, 7, 8, 9, 10, 11, 12, 13, 14, 15, TREES!" Point to the word fifteen trees and say, "FIFTEEN TREES!" Color away!!

"Fifteen Poster"

On the very next page, you have objects all about the number 15. Color the objects, cut them out, and paste them to complete your poster sheet!

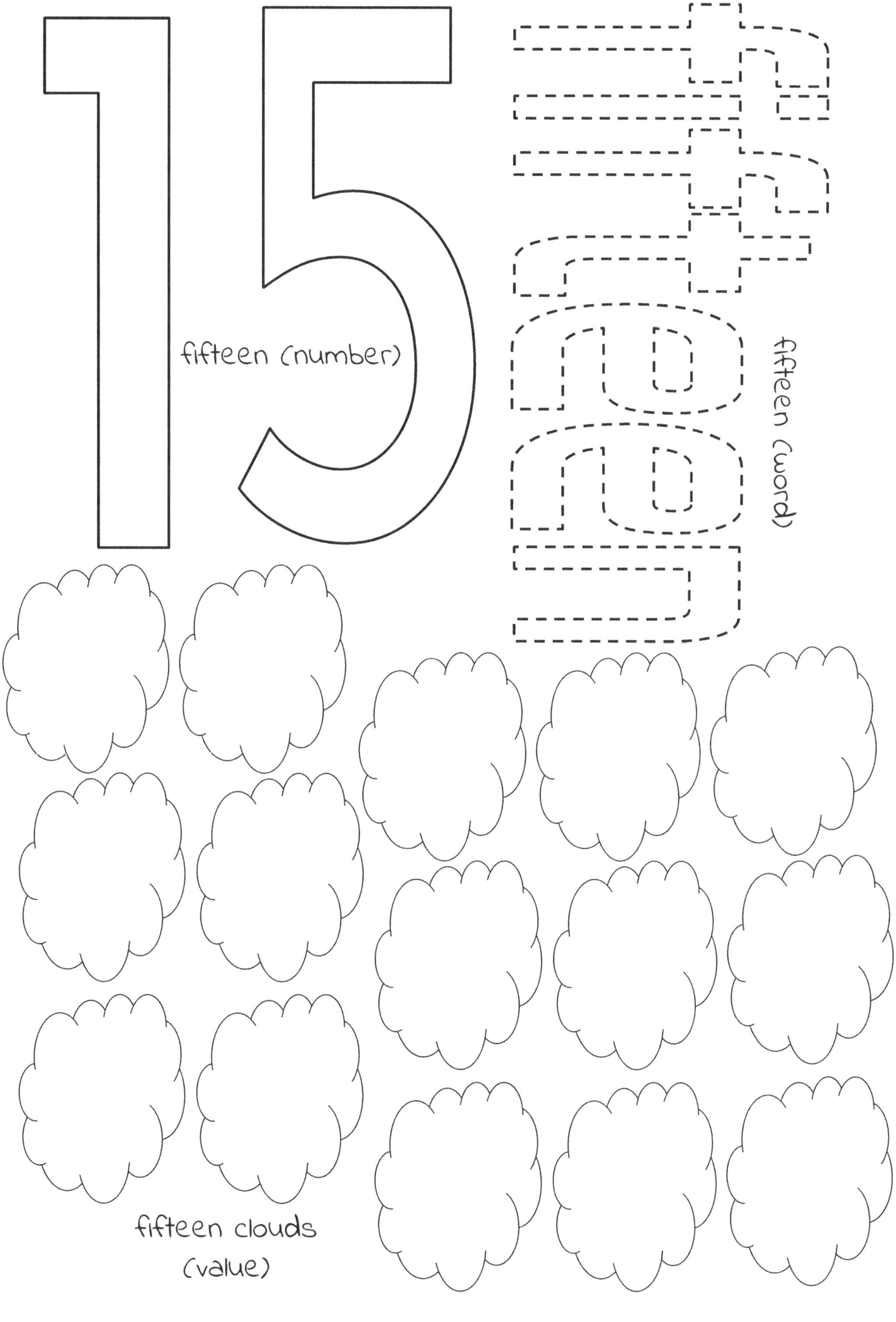
fifteen (number)
fifteen
fifteen (word)
fifteen clouds
(value)

Do you like to play in the falling leaves? I do too! Point and count the leaves, " 1, 2, 3, 4, 5, 6, 7, 8, 9, 10, 11, 12, 13, 14, 15, LEAVES!" Color your picture anyway you'd like!

16 "Sixteen"

sixteen

sixteen bird houses

What comes after the number 15? You guessed right, 16! Point to the number 16 and say, "SIXTEEN!" Point to the word sixteen, and say "SIXTEEN!" Point to the birdhouses and count, " 1, 2, 3, 4, 5, 6, 7, 8, 9, 10, 11, 12, 13, 14, 15, 16 birdhouses!" Point to the word 16 bird houses and say, "SIXTEEN BIRD HOUSES!"

"Sixteen Poster"

Color the objects on the next page. After coloring, cut them out, and paste them on this sheet. Now you have a nice 16's poster!

16
sixteen (number)
sixteen
sixteen(word)
sixteen triangles (value)

Have you ever went outside to see a field of birds? It's amazing! Count your birds with me, " 1, 2, 3, 4, 5, 6, 7, 8, 9, 10, 11, 12, 13, 14, 15, 16 BIRDS!" AWESOME!! Color your picture!

17 "seventeen"

seventeen

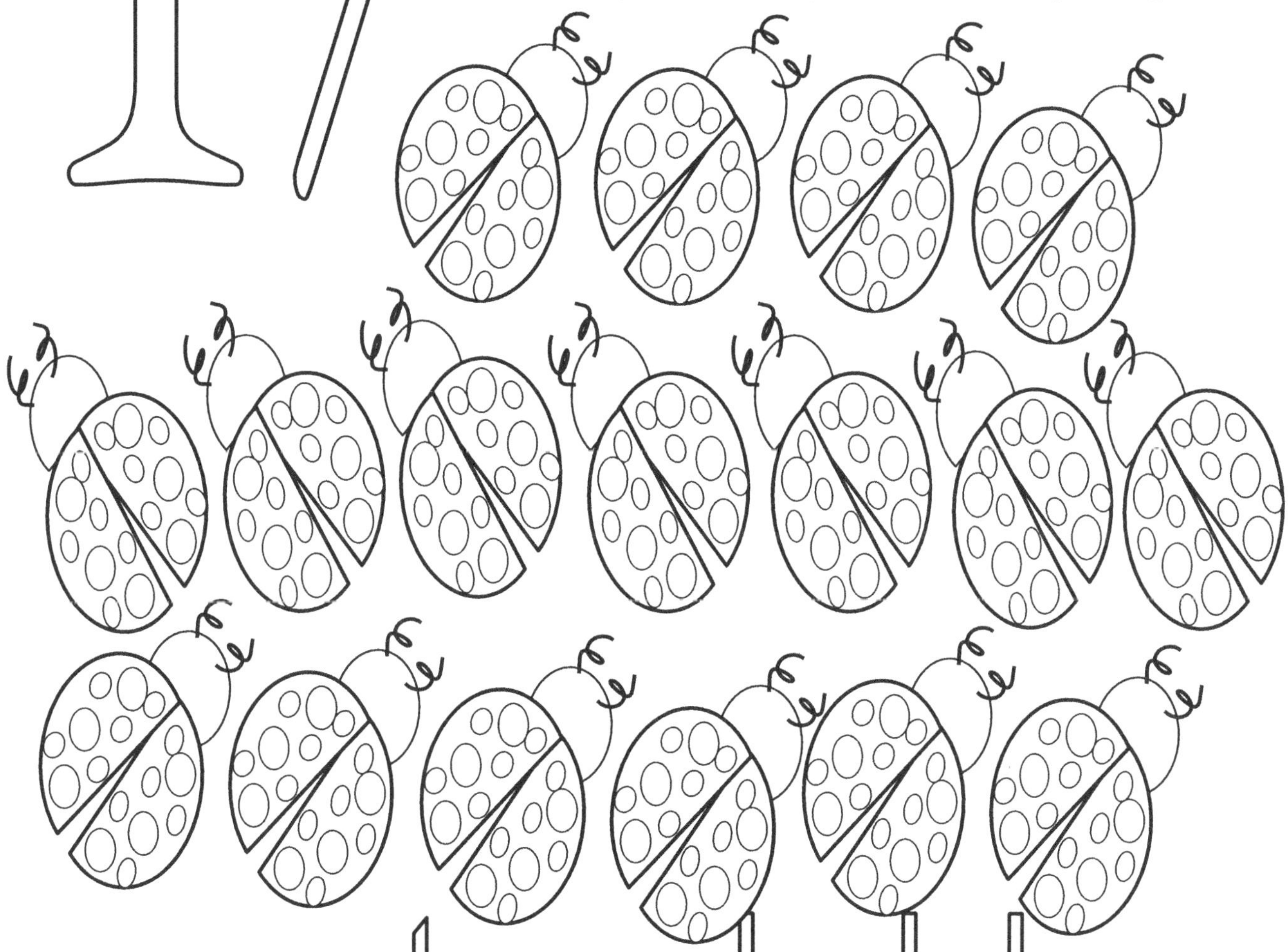

seventeen ladybugs

What comes after the number 16? YAY, 17! Point to the number 17 and say, "SEVENTEEN!" Point to the word seventeen and say, "SEVENTEEN!" Point to the ladybugs and count with me, " 1, 2, 3, 4, 5, 6, 7, 8, 9, 10, 11, 12, 13, 14, 15, 16, 17 ladybugs!" Color your picture!

"Seventeen Poster"

It's time to make this page finished. Color all the objects on the next page. Once you finish coloring, cut them out, and paste them on this page. Now you should have a filled poster worksheet!

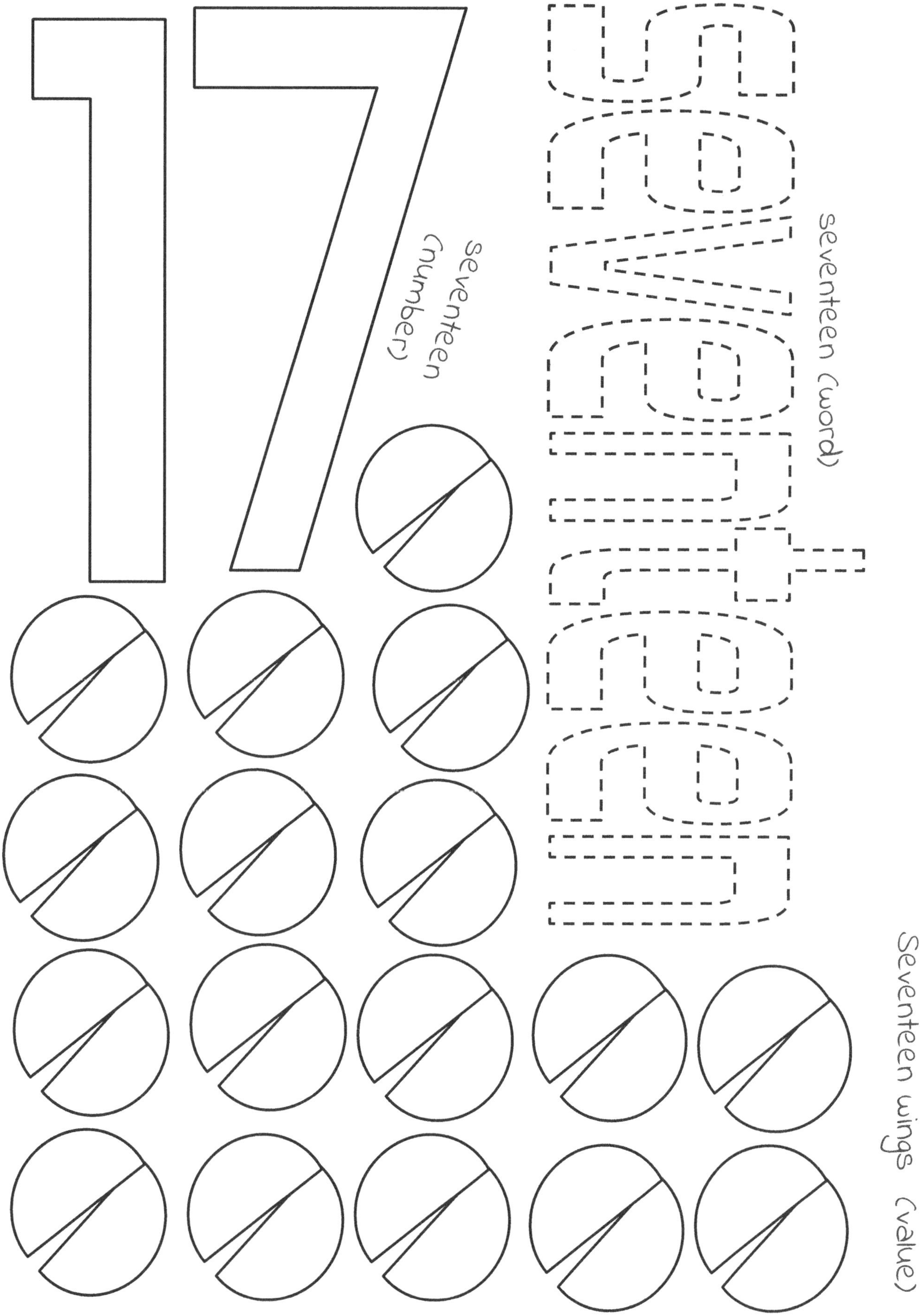
17
seventeen (number)
seventeen
seventeen (word)
Seventeen wings (value)

"Seventeen"

Can you count all of the leaves? Let's go! Say, " 1, 2, 3, 4, 5, 6, 7, 8, 9, 10, 11, 12, 13, 14, 15, 16, 17, LEAVES!" That was fun! Color your amazing picture!

18 "eighteen" eighteen

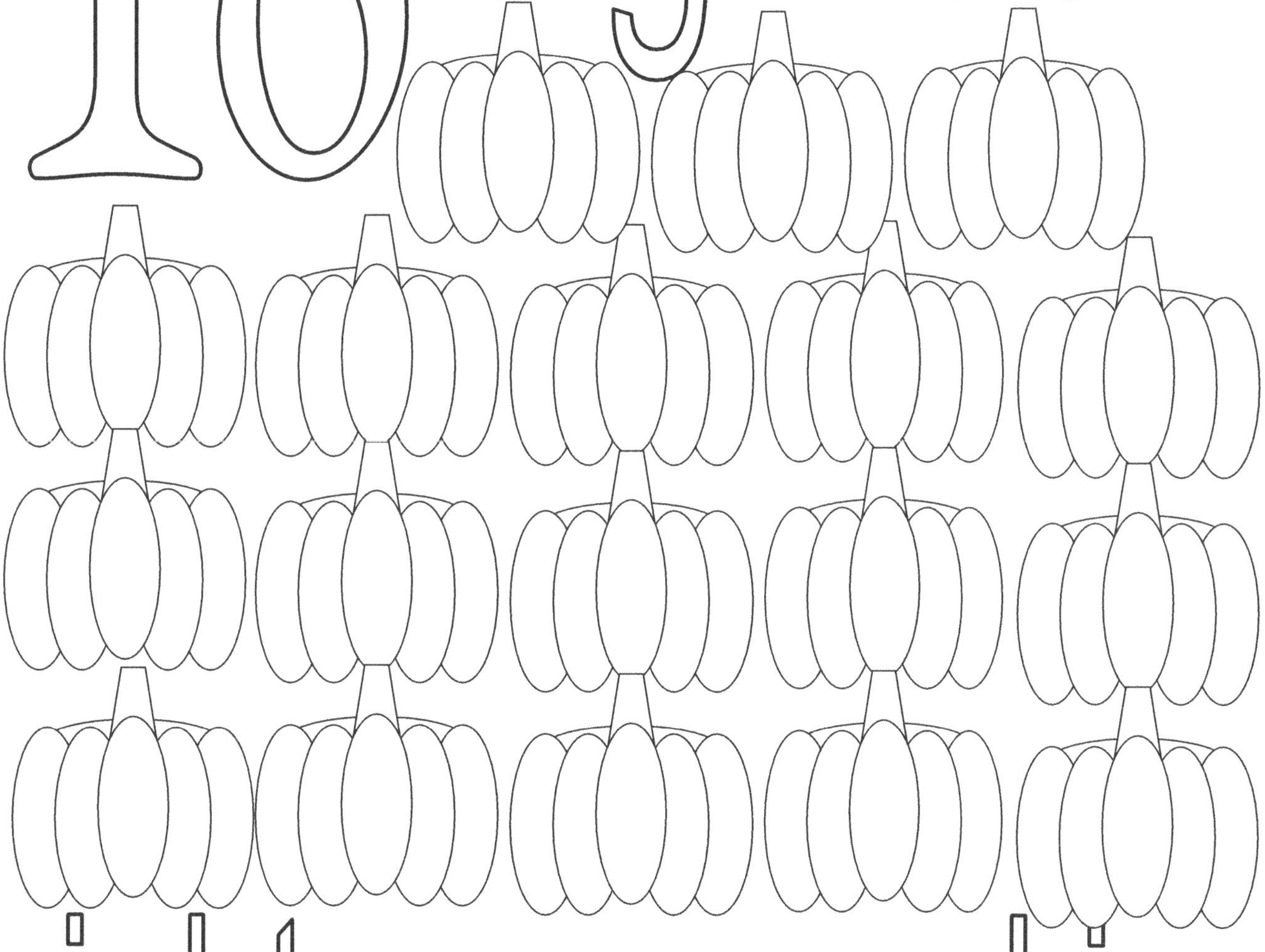

eighteen pumpkins

Do you know what comes after 17? If you said 18, then you're exactly right! Point to the number 18 and say, "EIGHTEEN!" Point to the word eighteen and say, "EIGHTEEN!" Point to the pumpkins and count them, " 1, 2, 3, 4, 5, 6, 7, 8, 9, 10, 11, 12, 13, 14, 15, 16, 17, 18, PUMPKINS!" Enjoy coloring your picture!

"Eighteen Poster"

On the next page you will find objects that talk about 18. Color your whole page, cut out all of the objects, and paste them on this page! You have completed your 18's poster!

eighteen (number)
eighteen (word)
eighteen trapezoids
(value)

Picking pumpkins is such a great Fall activity. Let's count the vines, " 1, 2, 3, 4, 5, 6, 7, 8, 9, 10, 11, 12, 13, 14, 15, 16, 17, 18, PUMPKIN VINES!" Have some more fun by coloring this page!

19 "Nineteen"

nineteen

nineteen paint cans

What comes after 18? That is correct! 19! You're on a roll! Point to the number 19 and say, "NINETEEN!" Point to the word nineteen and say "NINETEEN!" Point to the paint cans and count them, " 1, 2, 3, 4, 5, 6, 7, 8, 9, 10, 11, 12, 13, 14, 15, 16, 17, 18, 19, PAINT CANS!" Point to the word nineteen paint cans and say, " NINETEEN PAINT CANS! Color your picture!

"Nineteen Poster"

Color the objects on the following page. Cut out all of the objects and paste them here. You're poster looks great!

19
nineteen (number)
nineteen
Nineteen (word)
nineteen cans (value)

Any time we get to use paint it's always so much fun! Count the paint splatters with me, " 1, 2, 3, 4, 5, 6, 7, 8, 9, 10, 11, 12, 13, 14, 15, 16, 17, 18, 19, paint splatters!" Color your picture!

20 "twenty" twenty

twenty icecreams

What comes after 19? 20! Icecream, icecream, we love ice cream! Point to the number 20 and say "TWENTY!" Point to the word twenty and say, "TWENTY" Point to the ice creams and count, " 1, 2, 3, 4, 5, 6, 7, 8, 9, 10, 11, 12, 13, 14, 15, 16, 17, 18, 19, 20 ICE CREAMS!" Point to the word twenty ice creams and say " Twenty ice Creams!" Color your picture!

"Twenty Poster"

Flip over to the next page and you will find your coloring sheet. Color all the objects, then cut them out, and paste them to complete your 20's poster!

twenty (number)

twenty (word)

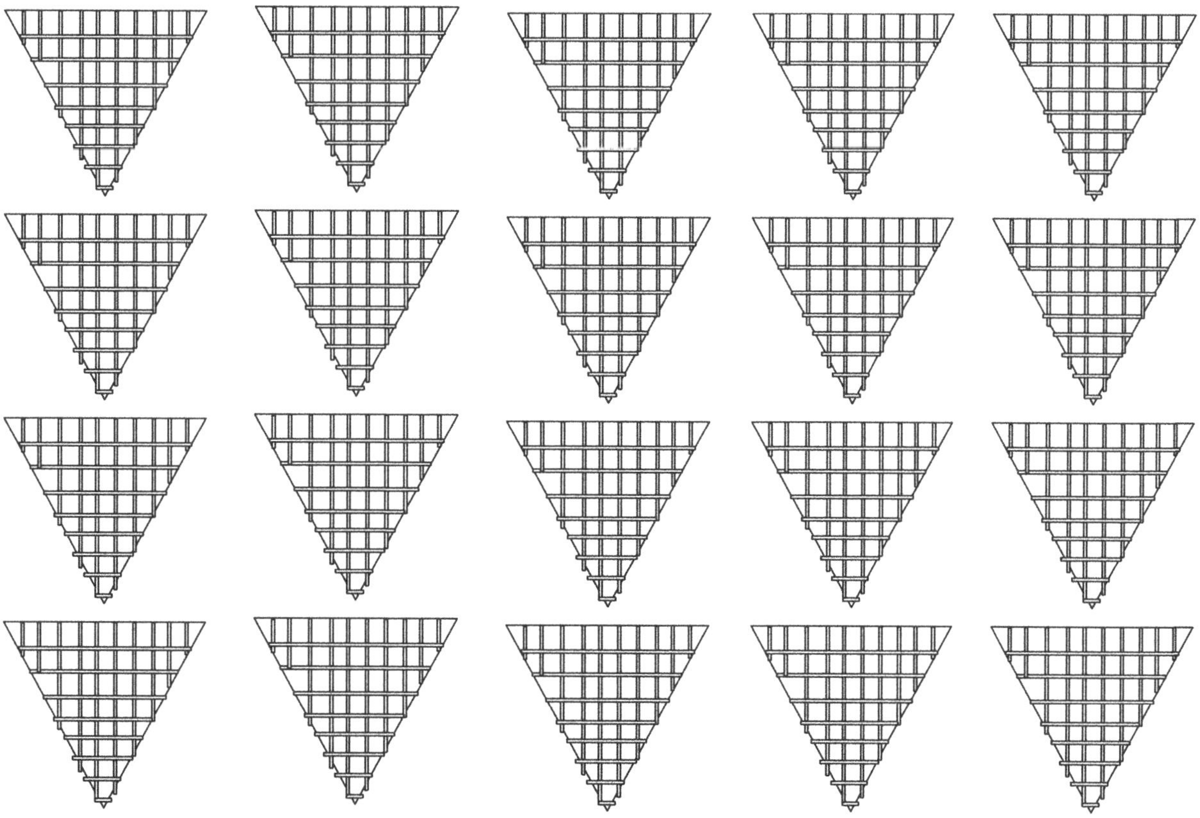

twenty cones (value)

Let's count all of these scoops of ice cream to see how many we have! Let's begin, " 1, 2, 3, 4, 5, 6, 7, 8, 9, 10, 11, 12, 13, 14, 15, 16, 17, 18, 19, 20, SCOOPS OF ICE CREAM!" Wow! That's plenty to share. Color your page.

"twenty-one"

21 twenty-one

twenty-one nuts

Comes after number 20? OMG, 21! Point to the number 21 and say, "TWENTY-ONE!" Point to the word twenty one and say, "TWENTY-ONE!" Point to the nuts and count with me, " 1, 2, 3, 4, 5, 6, 7, 8, 9, 10, 11, 12, 13, 14, 15, 16, 17, 18, 19, 20, 21, NUTS!" Point to the word twenty one nuts and say " TWENTY-ONE NUTS!" Color aways!

"Twenty-one Poster"

On the very next page you will find all the objects to go on this poster worksheet. Color the next page, cut out the objects, and paste them here. What a great looking poster!

21

Twenty one (number)

Twenty one hearts (value)

Twenty one (word)

twenty-one

Chocolate covered nuts and a beautiful card makes a great Valentine's Day gift! Let's count all of our cards to see how many we have! Count, " 1, 2, 3, 4, 5, 6, 7, 8, 9, 10, 11, 12, 13, 14, 15, 16, 17, 18, 19, 20, 21, CARDS!" Color your pretty picture!

22 "twenty-two"

twenty-two

twenty-two stars

What comes after the number 21? 22 is right! Point to the number 22 and say, "TWENTY TWO!" Point to the word twenty and say, "TWENTY TWO!" Point to the stars and count, " 1, 2, 3, 4, 5, 6, 7, 8, 9, 10, 11, 12, 13, 14, 15, 16, 17, 18, 19, 20, 21, 22, STARS!" Point to the word twenty two starts and say, " TWENTY TWO STARS!" Color you picture!

"Twenty-two Poster"

Turn to the page to color your objects. When you finish coloring your page, cut out the objects, and paste them here to complete your 22's poster sheet!

Twenty two (number)

twenty two wings (value)

"Twenty-Two"

Have you found a feather lying in the grass? We have collected plenty! Count the feathers with me, " 1, 2, 3, 4, 5, 6, 7, 8, 9, 10, 11, 12, 13, 14, 15, 16, 17, 18, 19, 20, 21, 22, FEATHERS!" Color your page!

23

"twenty-three"

twenty-three

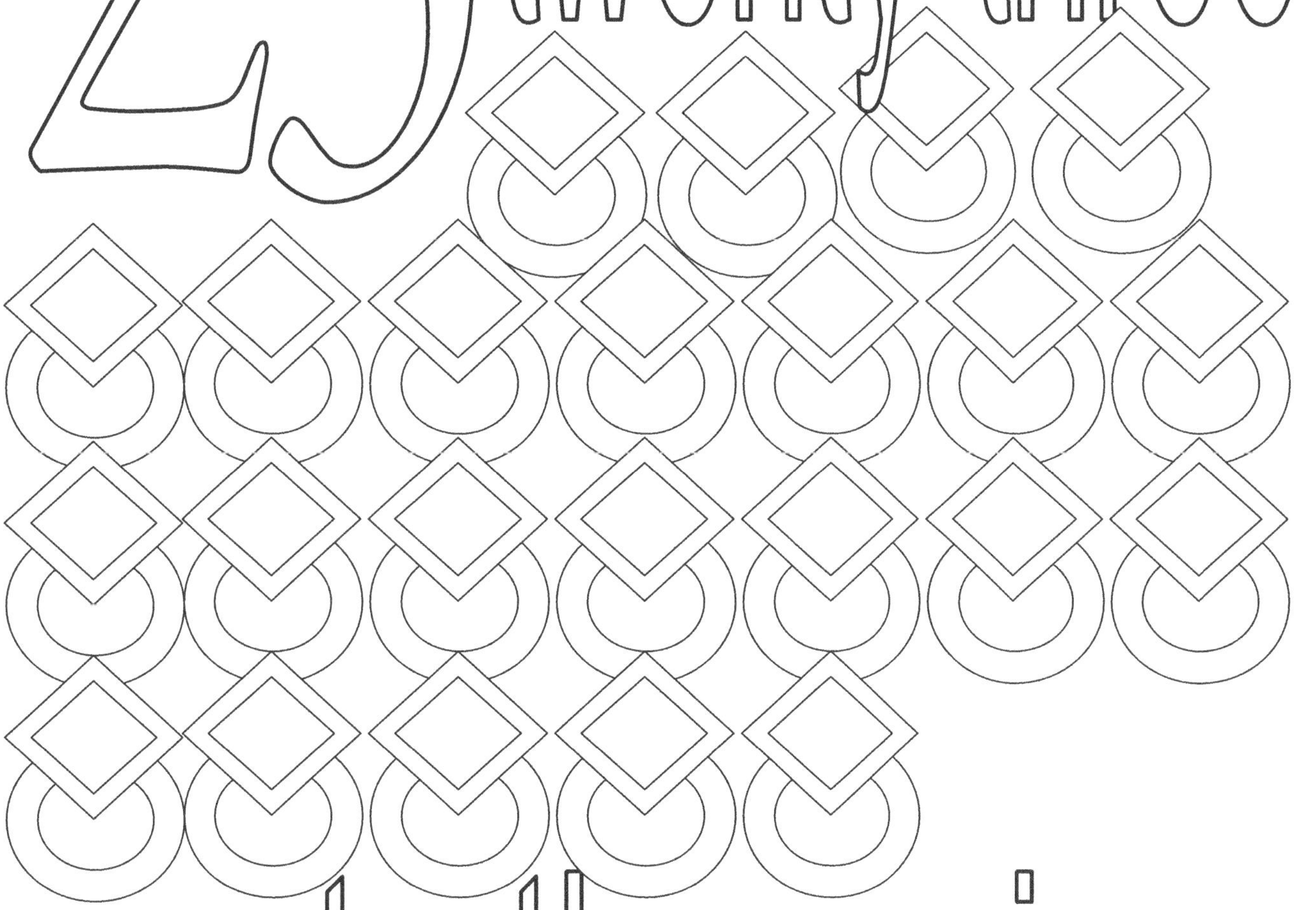

twenty-three rings

What comes after the number 22? You guessed right, 23! Point to the number 23 and say "TWENTY THREE!" Point to the word twenty three and say, "TWENTY THREE!" Point to the rings and count, " 1, 2, 3, 4, 5, 6, 7, 8, 9, 10, 11, 12, 13, 14, 15, 16, 17, 18, 19, 20, 21, 22, 23, RINGS!" Point to the word twenty three rings and say, " TWENTY THREE RINGS!" Coloring time!

"Twenty-three Poster"

This page won't stay blank for long. Turn to the next page and color your objects. After you finish coloring them, cut out the objects, and paste them on this page. Your poster worksheet should be nice and filled!

23
twenty three diamonds
(value)
twenty three (number)
twenty three (word)
twenty-three

"Twenty-three"

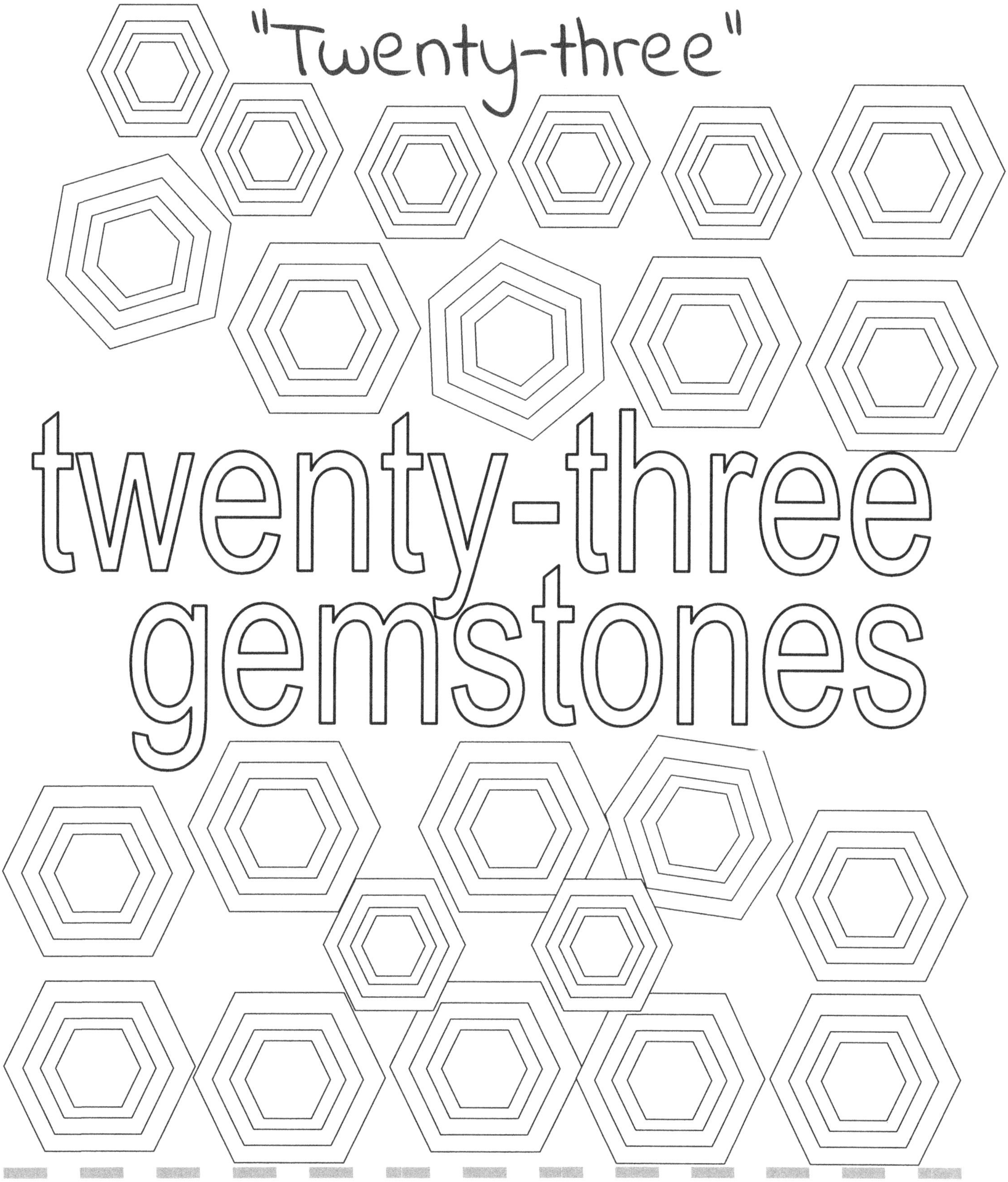

You are one rich kid! Count all of your gemstones, " 1, 2, 3, 4, 5, 6, 7, 8, 9, 10, 11, 12, 13, 14, 15, 16, 17, 18, 19, 20, 21, 22, 23, GEMSTONES!" Color your precious stones!

24

"Twenty-four"

twenty-four

twenty four grapes

Can you figure out what number comes after 23? That is right, 24! Point to number 24 and say "TWENTY FOUR!" Point to the word twenty four and say, "TWENTY FOUR!" Point to your grapes and count, " 1, 2, 3, 4, 5, 6, 7, 8, 9, 10, 11, 12, 13, 14, 15, 16, 17, 18, 19, 20, 21, 22,23,24, GRAPES!" Point to the word twenty four grapes and say, "TWENTY FOUR GRAPES!" Color your picture!

"Twenty-four Poster"

Flip over to the next page and color everything you see. Once you finish coloring, cut out all the objects, and paste them on this page to complete the poster! AND GO!

24
twenty four (number)
twenty four leaves (value)
Twenty four (word)
twentyfour

I love to eat yummy fruit! Count the bananas with me, " 1, 2, 3, 4, 5, 6, 7, 8, 9, 10, 11, 12, 13, 14, 15, 16, 17, 18, 19, 20, 21, 22, 23, 24, BANANAS!" Color your awesome picture!

What number comes after 24? Yes, 25! Point to number 25 and say, "TWENTY FIVE!" Point to the word twenty five and say, "TWENTY FIVE!" Point to the crayons and count, " 1, 2, 3, 4, 5, 6, 7, 8, 9, 10, 11, 12, 13, 14, 15, 16, 17, 18, 19, 20, 21, 22,23,24,25, CRAYONS!" Point to the word twenty five crayons and say, "TWENTY FIVE CRAYONS!" Color your picture!

"Twenty-five Poster"

This will soon be your 25 poster sheet. Color the objects on the next page. Cut out the objects and paste them on this page. Now you have an awesome 25's poster!

twenty five (number)

twenty five (word)

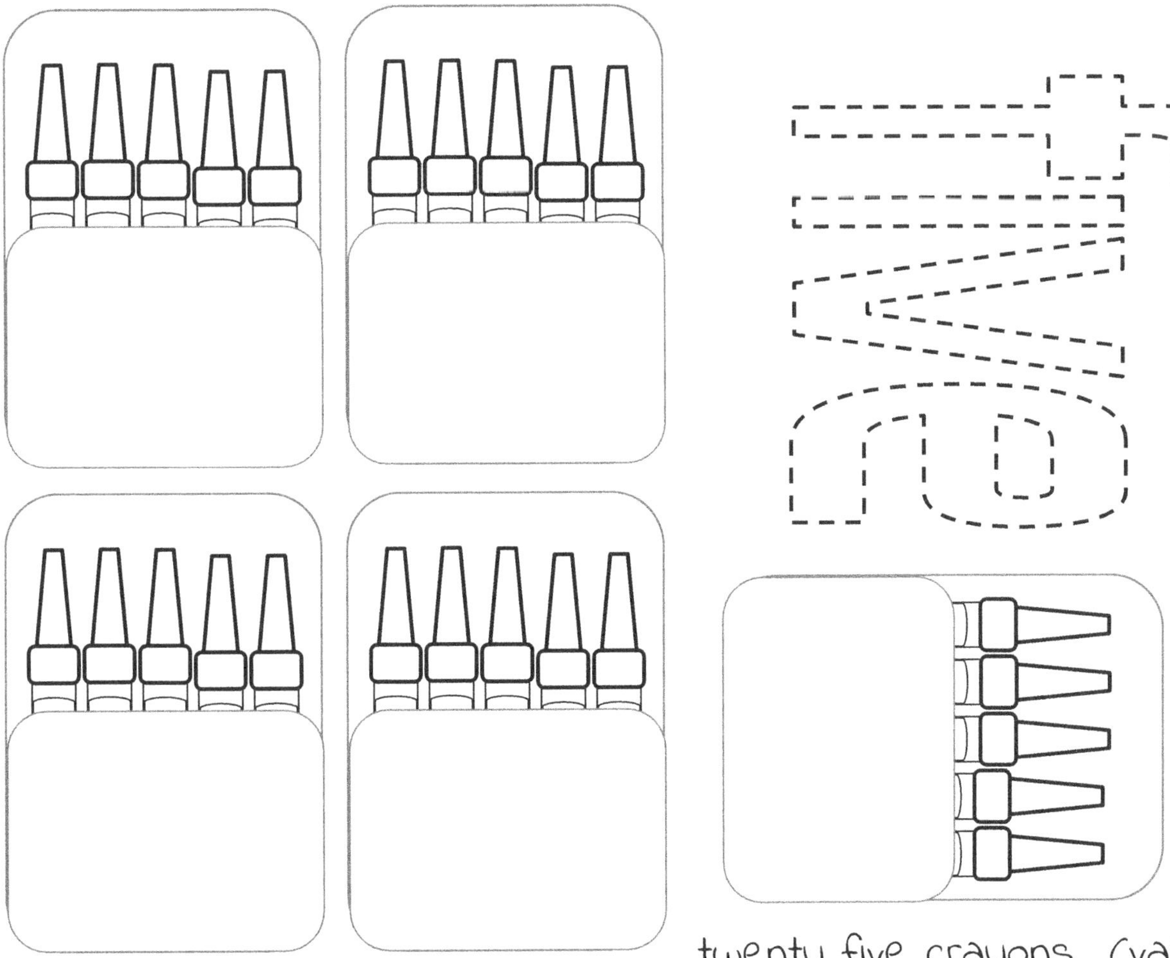

twenty five crayons (value)

twenty five sheets of paper

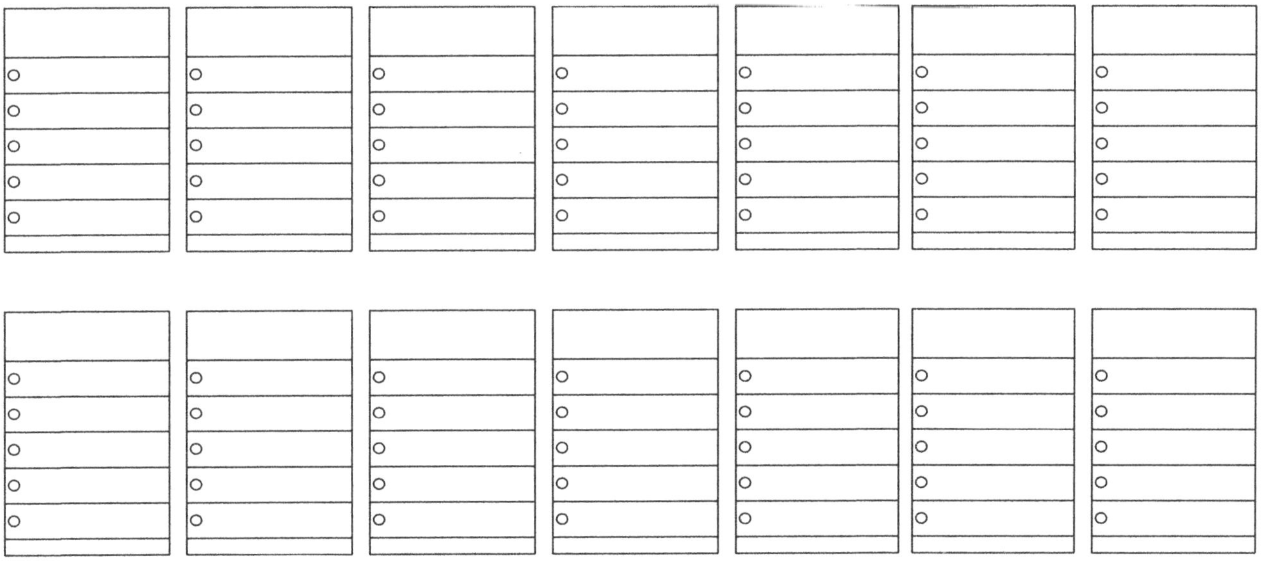

I love drawing, coloring and writing. Let's count our sheets of paper, "1, 2, 3, 4, 5, 6, 7, 8, 9, 10, 11, 12, 13, 14, 15, 16, 17, 18, 19, 20, 21, 22, 23,24,25, SHEETS OF PAPER!" Color your page!

Made in the USA
Monee, IL
08 November 2026

67380804R00116